西班牙建筑室内设计

Interior design of Spanish architecture

◆深圳视界文化传播有限公司 编

中国林业出版社

图书在版编目（C I P）数据

西班牙建筑室内设计 / 深圳视界文化传播有限公司 编.
-- 北京 : 中国林业出版社, 2018.7
ISBN 978-7-5038-9320-9

Ⅰ. ①西… Ⅱ. ①深… Ⅲ. ①室内装饰设计－西班牙
Ⅳ. ①TU238.2

中国版本图书馆CIP数据核字(2017)第254155号

编委会成员名单
策划制作：深圳视界文化传播有限公司（www.dvip-sz.com）
总 策 划：万绍东
编　　辑：丁　涵
装帧设计：潘如清
联系电话：0755-8283 4960

中国林业出版社 · 建筑分社
策　　划：纪　亮
责任编辑：纪　亮　王思源

出版：中国林业出版社
（100009 北京西城区德内大街刘海胡同7号）
网站：http://lycb.forestry.gov.cn
电话：（010）8314 3518
发行：中国林业出版社
印刷：北京利丰雅高长城印刷有限公司
版次：2018年7月第1版
印次：2018年7月第1次
开本：1/16
印张：20
字数：300千字
定价：428.00元

Amaro Sánchez de Moya is a Spainish architect and interior designer. Early artistic vocation as a painter and architect, Amaro matured academically in Seville (Spain); Venice (Italy) and Paris (France), ultimately, this wealth of knowledge and experience allowed Amaro to develop his unique sensibility as an interior designer.

In 2007, Amaro opened his studio in Seville, where he develops a wide range of residential, commercial interior design projects. A true world citizen, Amaro has worked internationally in a variety of cities, such as Paris, Saint-Tropez, Venice, Lisbon, Madrid, and Seville. Amaro has been featured in numerous print and online publications and he his works leave an enduring impression on design enthusiasts.

阿马罗·桑切斯·莫亚是一名西班牙建筑师及室内设计师。他早期曾做过油画家和建筑师，在西班牙塞维利亚，意大利威尼斯和法国巴黎有过求学经历，而从中获得的知识和经验发掘了他自身的才能，最终成为了一名室内设计师。

阿马罗于2007年在塞维利亚创办了自己的公司，设计了许多住宅和商业案例。作为名副其实的世界公民，他设计的案例遍布世界各大城市，比如巴黎，圣特罗佩，威尼斯，里斯本，马德里和塞维利亚。他的作品一直被刊登在众多印刷和在线出版物上，对设计爱好者们产生了重大影响。

CONTENTS 目录

A MEDITERRANEAN CRUISE

地中海之旅

设计公司
DESIGN COMPANY
MELIAN RANDOLPH S.L.

设计师
DESIGNERS
Victoria and Sylvia Melian Randolph

摄影师
PHOTOGRAPHER
Martin Garcia Perez for Nuevo Estilo/ Hearst

The house was built in the 90's as a rationalistic beach bungalow on the coast of Ibiza overlooking the Mediterranean by the architects Planas and Torres. The project required respecting the original structure in glass and steel with white marble floors, but reconverting it into something more comfortable with bedrooms, bathrooms and air conditioning, etc. The client is an international bachelor, who enjoys sports and receiving guest, so he requested large open spaces where he could receive indoors and outdoors.

始建于90年代，这个房子是由建筑师普拉纳斯和托雷斯建造的一个偏理性主义的平房，位于伊比沙岛海岸，俯瞰地中海。这个项目要求尊重玻璃、钢铁和白色大理石地板的原始结构，在此基础上，把它改造成有卧室、浴室和空调等的舒适套房。客户是个国际学者，喜欢运动和待人接客，因此他要求有更大的户内外空间招待客人。

The bathrooms and kitchen also required modernizing, and the lighting was renewed to adapt to the new distribution. The master bedroom at the top of the stairs was very dramatically open the outside, it now looks towards the Mediterranean instead of into the house. The basement now has an enclosed one car garage, a gymnasium and an extra guest bedroom and bath.

浴室和厨房也要现代化的设计，照明设施也重新布置以适应新的布局。楼道顶上的主卧面向户外，看起来像是走进了地中海而不是走进了房内。地下室现在增加了一个封闭的车库、体育馆、客房和浴室。

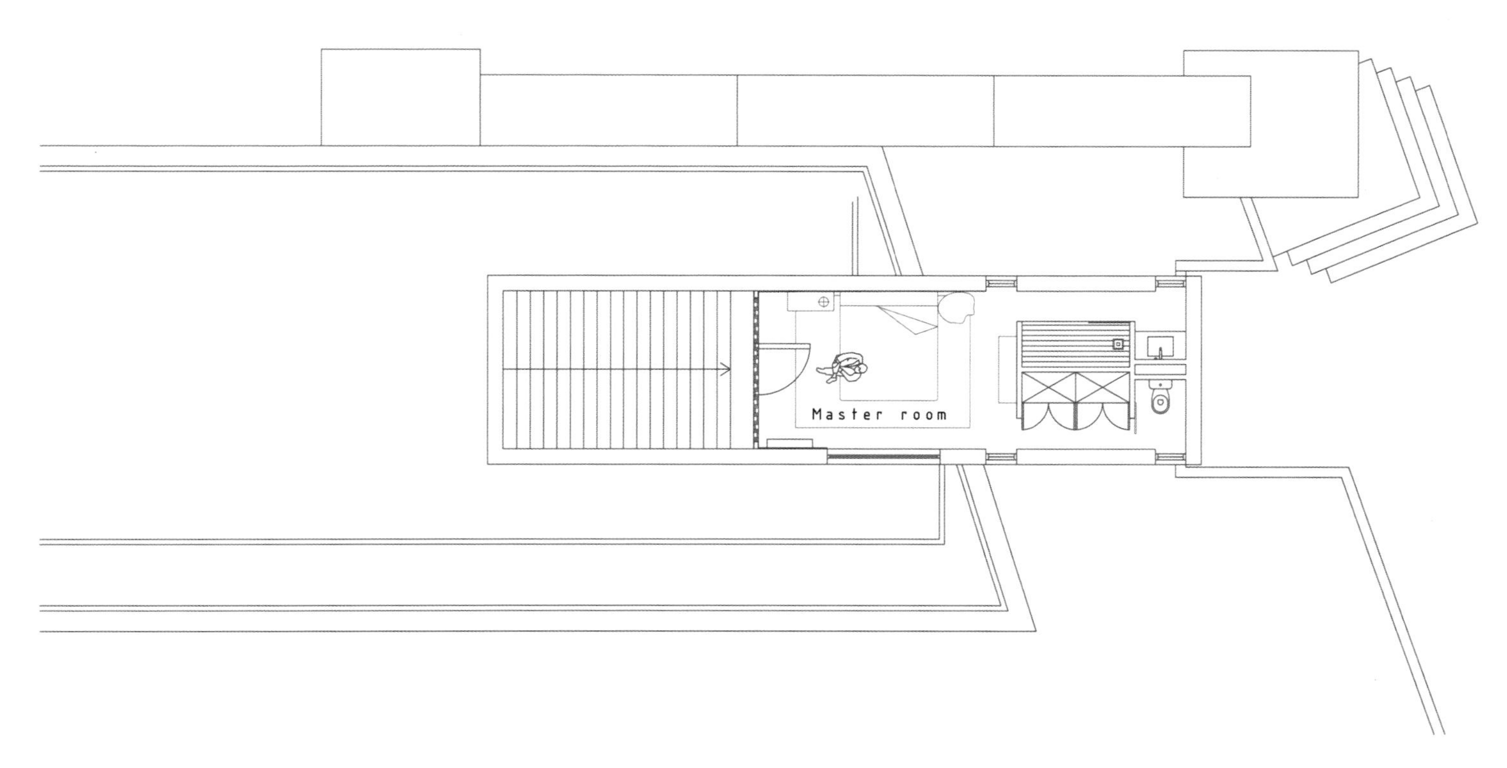

First floor_Casa Rufus_Cala Carbó_Ibiza

A BRIGHT AND SNOWY SPACE

亮白空间

设计公司
DESIGN COMPANY
Manolo Yllera

设计师
DESIGNER
Manolo Yllera

摄影师
PHOTOGRAPHER
Manolo Yllera

This is a nonsense story with a happy ending. Here Manolo Yllera failed all his goals of saving space, money and practicality. He wanted an open big empty space, the classical loft, so he could do his photography, moving the flash and spotlights easily. At the same time he wanted a home. He had a clear dual purpose. It was hard to find this space in the center of Madrid. It took him one year to find this place, without columns in the middle of the room, with 4.5 meters high, lateral light, and clean separation between spaces: the house and the photography set.

JUMBO

这是一个有着美好结局的荒诞故事。马诺洛·伊叶拉的目标是节约空间、金钱和实用性，然而这些都没有实现。他想要一个很大的开放空间，就像传统阁楼那般，这样他就可以摄影，轻松地挪动闪光灯、聚光灯等设备。同时，他也想要一个家。他有明确的双重目的。然而在马德里市中心很难找到这样的地方。他花费了一年时间，才找到了这么一所房子，房子高4.5米，中央没有圆柱，横向光照，并且可以清晰地区分居所和摄影空间。

The initial idea was to preserve the photography set as clean and empty as possible, to suit the needs of different productions. The problem was that his passion for decorating was gradually gaining ground.

First a chair reupholstered with sacks, then a Louis XV sofá eighteenth century fascinated him, then more chairs, carpets, and many other lovely things to the point of conquering almost the entire space of the set. Every time they needed to do a shooting for a client, they had to move most of the furniture to the street, or the room, or anywhere. A crazy impractical way of working. The decoration was doing more difficult the photography business till a point Manolo's assistants could not believe it. The surprise was that the decor brought his own business and clientele. Many people were interested in renting the place to do their own shootings, and gradually his space was the scene for video clips (more than 10 films and commercials, etc). The kitchen has been filmed for some show cookings, in the sofas some models have dance for Womans Secret, in the chairs some electric guitars have been held for music covers, like the armchair with fabric bag that has been published more than 8 times.

最初的想法是尽可能干净、宽敞地保留摄影空间，以适应不同产品的需求。问题是他对装修的热情却与日俱增。

起初他用布袋重装椅面，后来又对18世纪路易十五的沙发着迷，接着更多椅子、地毯和其他可爱的饰物一起，几乎占领了整个摄影空间。每次要为客户摄影的时候，都需要把这些家具搬到街边、屋内或者任何地方。这是一种疯狂的不切实际的工作模式。这种装修一直令摄影生意不佳，直到发生一件事令情况转变，连马诺洛的助理们都不能相信，这装修本身招揽了生意和客户。很多人对租用这个地方感兴趣，做他们自己的摄影，渐渐地，他的空间成为了视频剪辑的场景（有超过10部电影及商业宣传片在此取景）。厨房被用来拍摄、展示厨艺，一些模特在沙发上为“女人的秘密”翩翩起舞，椅子上的电吉他用来做音乐专辑封面，布袋装饰的扶手椅被出版过8次以上。

The style is a mix of furniture found on the street, classic furniture, some extreme design like some Marteen Baas pieces, furniture inherited from Manolo's grandmother painted in a radical way to participate in the new spirit of the space, Atlas Morocco carpets, industrial furniture, photography lights used daily in his work, some pieces from friend artist, paintings on the wall draws by friends too, etc.. This space is a mirror of Manolo's life and his profession, both have converged together in the same spot.

这里的风格是街头家具的混搭，古典家具、一些像马丁巴斯般偏激的物品、马诺洛祖母遗留的家具被漆以夸张激进的色彩来适应空间的新风貌、摩洛哥的阿特拉斯地毯、工业风家具、他日常工作使用的摄影灯具、艺术家朋友的物件和墙上朋友们绘制的油画等等。马诺洛这些生活和职业的反映，都融合在这个空间里。

ART APARTMENT

艺术之家

设计公司 DESIGN COMPANY	设计师 DESIGNERS	摄影师 PHOTOGRAPHER
recdi8 studio	Alex Baeza, Albert Vuidez, Norbert Frei	Roberto Ruiz

On one hand, the design concept is based on a 'dialog' with the history of the apartment (built 1905), but in terms of just looking back, and not deliberately imitating the original appearance. On the other hand, the interior is also based on the extensive art collection of the apartment's owners. The design of interiors is complementary to the numerous art pieces from the 16th, 17th and 2oth century.

本案设计理念一方面基于这间公寓（1905年建成）的历史"对话"，追忆却并不刻意模仿假想中的原貌。另一方面，室内空间要能摆放主人繁多的艺术收藏。这些空间的设计需与16世纪、17世纪及20世纪重要的艺术品互相匹配。

The style is classic decorative and contemporary functional at the same time. The designer has emphasized on curved lines, light neutral color palette mixed with highlights of gold, yellow, purple and black & white. The used materials are basically: brass, oak wood, lacquered wood, velvet, ceramic, leather, glass, marble.

在这个项目中，经典装饰风格与当代功能共存。设计师强调曲线，也强调浅中性色彩与金色、黄色、紫色及黑白等色彩的搭配。本案运用到的材质主要有：黄铜、橡木、漆板、丝绒、陶瓷、皮革、玻璃以及大理石。

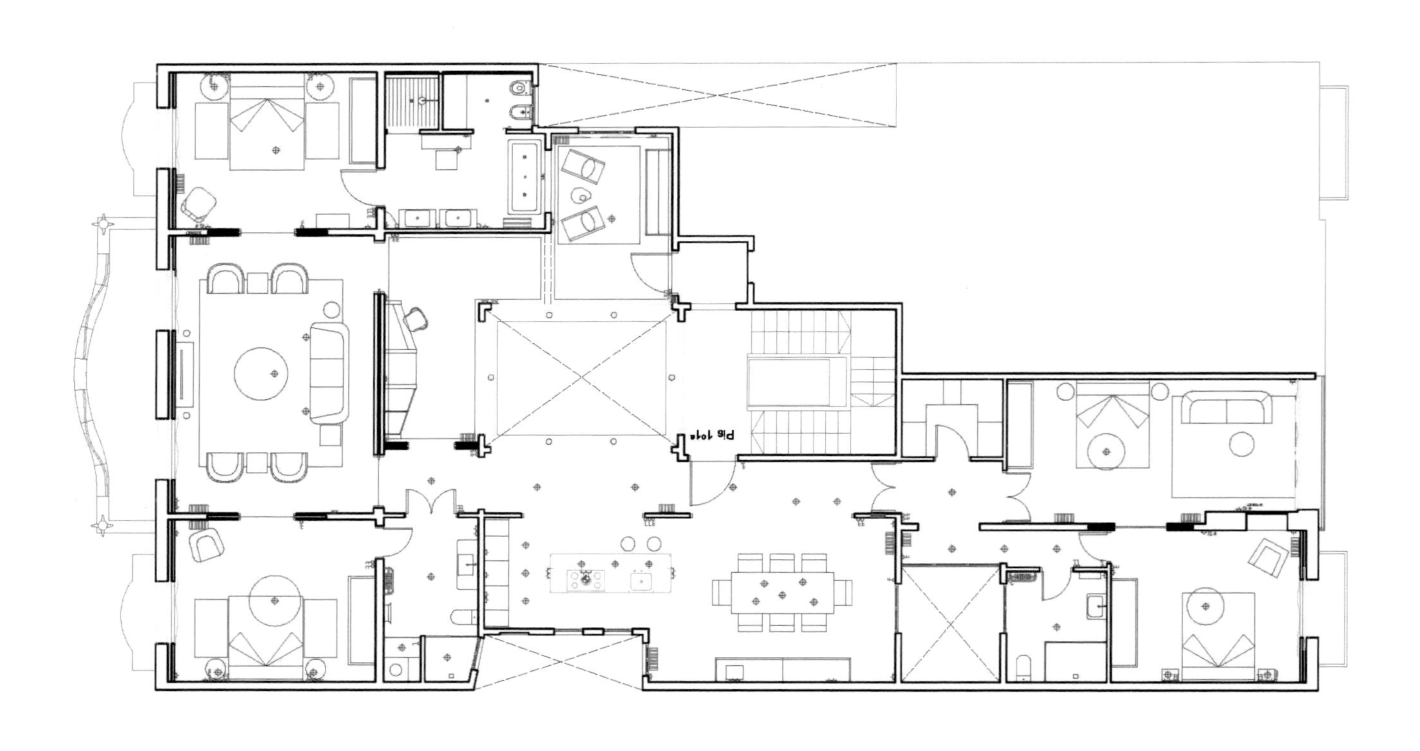

EXIGEZ
UN
PEUREUX
Un Peureux
PEUREUX FILS

Glamour

SPAIN LIVING · MADRID

MODERN CLASSIC
摩登经典

设计公司	设计师	摄影师
DESIGN COMPANY	**DESIGNER**	**PHOTOGRAPHER**
Estefanía Carrero Estudio	Estefanía Carrero	Pablo Sarabia for Nuevo Estilo / Hearst Magazines

This residence is in a 1800´s manorial building located in the Barrio de las Letras of Madrid, authored by architect Juan de Madrazo, brother of the famous painter. The project seeks to recover the formal and stylistic aspects of the property, which had been lost over the years. To achieve this goal, the designer restored the ancient woodwork (or rebuilt it where it had been replaced), recovered the original height of ceilings and rescued former scheme distribution, always being respectful with the style of the building itself and implying that the space had never changed. To balance and maximize this idea, the designer chose 1900´s furniture along with other own designs, light colors to enhance daylighting coming from large windows, ocher tones in the upholstery for a more comfortable feeling and a 4 meters tree to bring in the vegetation outside the street and achieve more depth to the space.

这所住宅位于马德里的一座19世纪庄园内，由一个知名画家的兄弟胡安·马德拉佐建造。本案力图在其历经沧桑之后恢复其面貌。为达到这一目的，设计师修复了古老的木工（或者在更替处重新建立），恢复了天花原始高度，还原了之前的设计布局。设计师对该建筑自身风格一直心存敬畏，并暗示自己这个空间从未被改造过，因此选用了20世纪有独特风格的家具。浅色调的运用增加了从窗外引入的采光，赭色系的家居衬垫营造更舒适的氛围，4米高的大树将街边植被带进室内，为空间增添了深度。

TYLE BY SALADINO

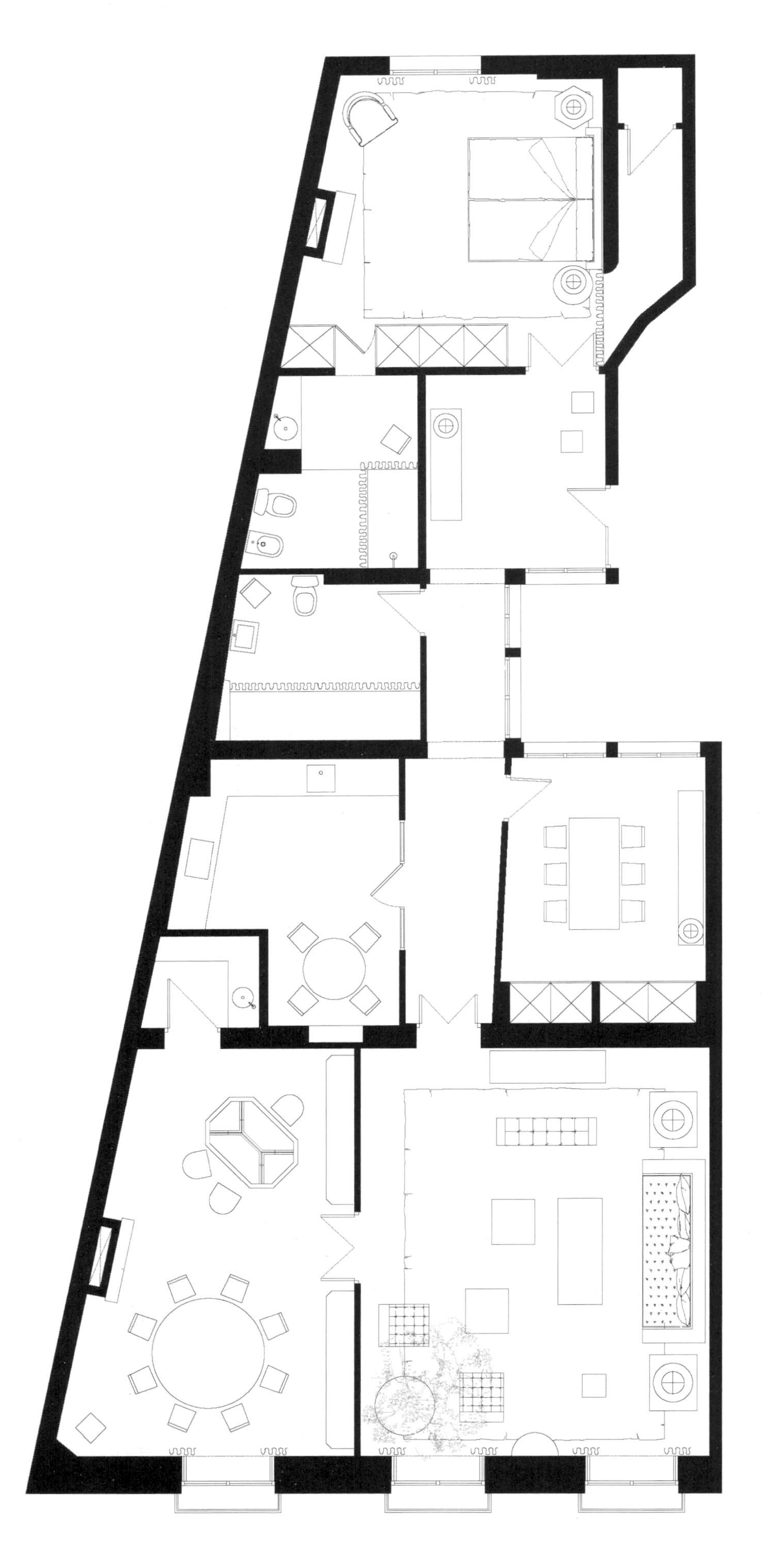

JAMIE
MILLER'S
Styles

AN INVITING HOME

魅力之家

设计公司
DESIGN COMPANY
Blanca Fabre Studio

设计师
DESIGNER
Blanca Fabre

摄影师
PHOTOGRAPHERS
Pablo Sarabia / Rafa Diéguez
Nuevo Estilo / Hearst

Comfortable contemporary style with a touch of modern. Big spaces divided by fixed screens and glass doors to have a whole view of the distributions. Exclusive unique furniture made by order for each project mixed with mid century modern and important pieces of art. Architectural section highly developed to meet the client expectations like distributions, materials, lighting design, ceilings, custom designed doors, moldings and baseboards, creating a personalized set that harmonizes with decoration.

本案为带着一丝时髦触感的现代住宅。大型空间借由固定屏风及玻璃门分隔而成，以达成布局的整体观感。专属独一无二的家具都按照业主需求逐个打造，混搭着中世纪珍贵的艺术品。采用明确的分区，可满足屋主对布局、材料、灯光、天花、定制门、线脚及脚板等方面的期望，从而打造一个与装修和谐共生的个性化空间。

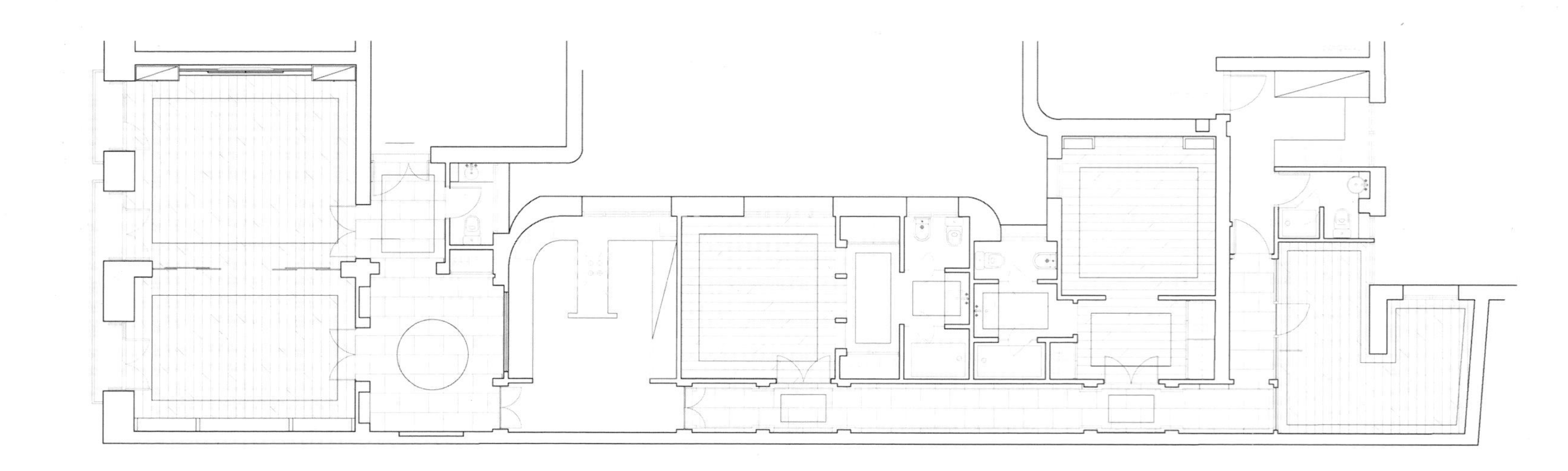

Architectural work made with a neutral palette painting and warm materials like oak, walnut, plaster moldings in contrast with antique mirrors on the walls, marble floors with pattern design, steel/glass screens and ceiling lights in art deco style. Kitchens specially designed with custom made lacquered furniture to provide the same colors in the whole project. About decoration, all furniture is chosen or designed in different and exclusive materials such as brass, steel, marble, resin, shagreen, mirror or plaster among others to create a luxury mix. Designer also chooses a large selection of international fabrics for upholstering and curtains and custom made designed rugs made with cotton and silk.

本案采用中性配色，选用温和材质，如橡木、胡桃木、塑料模型等，与装饰艺术风格的古董镜、大理石花纹地板、钢/玻璃屏风及天花顶灯形成对比。厨房采用定制漆面家具这一特殊设计，展现出与整体方案一致的色彩。所有家具都选用了不同材料，如黄铜、钢材、大理石、树脂、鲨革、镜面及塑料等，以此打造一种混搭的奢华风。设计师还在窗帘及纯棉丝绸定制地毯上运用了大量精选的国际面料。

Nuevo Estilo

SPAIN LIVING · SAN LORENZO DE EL ESCORIAL, MADRID

ELEGANT COUNTRY HOUSE

乡村雅舍

设计公司

DESIGN COMPANY

MASFOTOGENICA INTERIORISMO
www.masfotogenica.com
www.lexingtoncompany.com

设计师

DESIGNERS

PILAR MOLINA NAVARRO & CAROLINA VERDUGO

摄影师

PHOTOGRAPHER

CARLOS YAGÜE RIVERA for INTERIORES Magazine (Prisma Publicaciones)

Owners of international brand LEXINGTON are villages and horses lovers. The area of San Lorenzo de El Escorial, Madrid is the place where they combine work with home: the head office of the company is next to mountains, just 10 minutes by car. This is a big and strange stone house prepared with all amenities for a big family who love to enjoy life together.

国际品牌列克星敦的创立者钟情于乡村和马。所以西班牙的圣洛伦索·德埃尔埃斯科里亚尔地区就成了他们把工作和生活结合起来的地方，公司总部靠近山，大约10分钟车程。这个房子是一个很大很奇怪的石头房子，但设备齐全，住着一起享受美好生活的大家庭。

Eat
Drink
BE
MERRY
SNACKS

EXINGTON

This awesome country house is located in the mountains next to Madrid City, capital of Spain, just 40 minutes by car. The house, surrounded by meadows with grazing horses, has three floors. In the first floor with access from the ground, there are 3 bedrooms, one of which is the master bedroom with a bathroom and a dressing room en suite. There are also 2 more bathrooms, a living room, a kitchen and an office. In the basement, there are two guest bedrooms and spaces for storage and laundry. In the top floor there are 2 attic rooms with a bathroom and a dressing room.

Decorators Pili Molina and Carolina Verdugo, from Masfotogenica.com, are responsible for main decoration of this house. All fabrics, household linen and beddings were acquired from the family brand LEXINGTON. With an informal and very welcoming style skillfully blending classic pieces with accurate touches of contemporary art, the house is completely designed to be enjoyed with family and friends.

这个乡村住宅在山里，离西班牙的首都马德里只有40分钟车程。这个房子有三层，周围都是牧马草地。一楼从地面直接进入，有三个卧室，其中一个卧室是主卧自带浴室和更衣室，此外，还有两个洗手间、起居室、厨房和办公室。地下室有两个客房、储存室和洗衣房。顶楼有两个阁楼房间、浴室和更衣室。

这个房子的装饰主要是由Masfotogenica.com设计公司的装饰家皮利·莫利纳和卡罗林·贝尔杜戈完成的。所有的纺织品，家用亚麻制品和床上用品都来自于列克星敦。这个房子的设计风格是巧妙地将古典元素与现代艺术相结合而形成的一种非正式却很受欢迎的风格，家人和朋友在这里享受着生活。

SPACE FOR
MINI
I'D RATHER
STAY IN BED
Club

I'D RATHER
STAY IN BED
Club

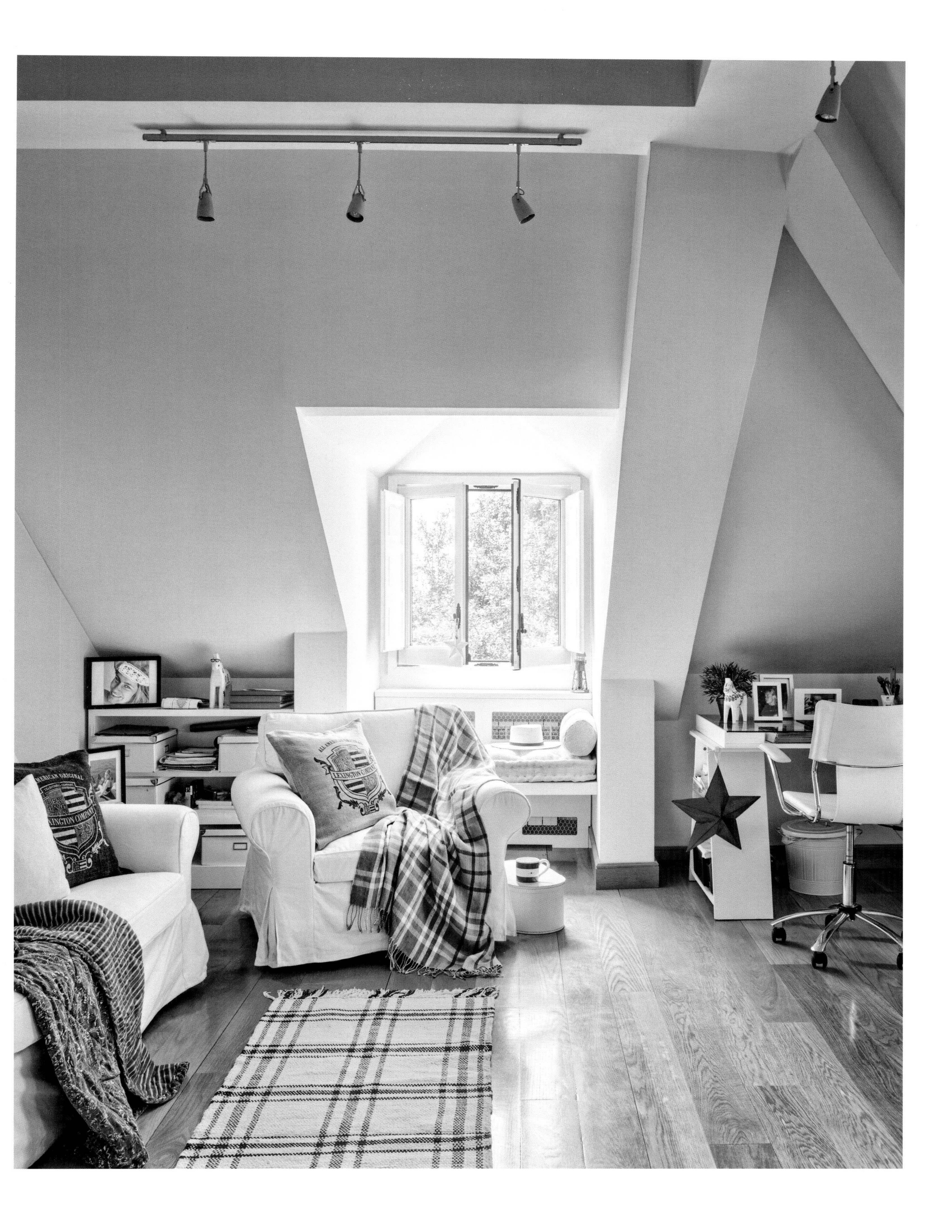
LEXINGTON

The house was purchased years ago from a dealer, and has respected some of the main existing elements such as the large fireplace and the stone floor of the room. The rest of the house has been retouched over the years. The house is painted in soft colors to make it brighter, but has also used the paper on the walls as a decorative element.
It is a mix of spanish tradictional style.Very strong walls made by stone and an eclectic spirit are very common in the style of Carolina Verdugo and Pili Molina for interior design. They choose what they love and make it work together. The love to create homes ready to be lived, and they like to be practical as well as aesthetic. "A very beautiful house is the one that is comfortable to live in. If it is just beautiful but you dont want to live in, then it is not 100% perfect" says Pili and Carolina. "Our first step always is to know exactly what the owners expect for the house, then we follow those ideas to transform them into something real. At the end a new home with it own soul is what we create. Then, if they are happy, we are happy. For us decoration it is not just a job, it is our vocation".

这个房子是多年前从一个经销商手里买来的，但却保留了一些之前的元素，比如大壁炉和房间里的石头地板。房子的其他部分多年来有所修整。为了使房子看起来更明亮，就将它粉刷成了柔和的颜色，此外，墙上也用了壁纸作为装饰元素。

这是一种混合的传统西班牙风格。坚硬的石头墙壁和折衷主义精神在卡罗林·贝尔杜戈和皮利·莫利纳的室内设计风格中是很常见的。他们选择自己喜欢的装饰品并把它们结合起来。他们喜欢设计既实用又美观，还可以随时居住的房子。“一个漂亮的房子应该是适合居住的房子，如果它只是漂亮，而你不想住在这里，就不是百分百的完美。”皮利和卡罗林说。“我们的首要任务是了解屋主对这个房子的期望，然后根据这些期望付诸行动，最终创造一个有自己灵魂的新家。屋主高兴了，我们也就高兴了。对我们来说，装饰不仅仅是一份工作，而是我们的使命。”

ECLECTIC AND ELEGANT

折中与优雅

设计公司

DESIGN COMPANY

Soledad Suárez de Lezo Interior Design

设计师

DESIGNER

Soledad Suárez de Lezo

摄影师

PHOTOGRAPHER

Jaime Boira

ROYAL OAK
ARTE
MILES REDD THE BIG BOOK OF CHIC

The team had to transform this very worn out and dated 1910s flat into a family home but never losing the sense of style and sophistication. Their aim was to achieve an eclectic but still elegant style while enhancing the beautiful architectural features the apartment had. It was a great challenge keeping a conservative criteria while updating the needs and style, but well worth it.

设计团队需要将这间1910年左右破旧不堪的公寓打造成一个温馨的家，并且不失其旧日风貌。他们的目标是在提升公寓原有建筑特色的同时，呈现一种折衷而高雅的风格。在满足新需求和打造新风尚的同时，坚守传统的准则是很大的挑战，但值得一试。

ARCHER
FREDERICK FORSYTH

This project came together really easily and quickly. One thing soon lead to another. The starting point were the architectural features and the entrance hall which they wanted to turn from gloomy into a welcoming kind of conservatory with marble floors, iron and beveled glass doors and plants.

本案设计过程简单且迅速，并且进行有序。首先进行改造的是建筑特色以及入口大厅，客户希望将其从一个幽暗空间改造成由大理石地面、铁艺玻璃门及绿植构成的魅力空间。

What is very unique about this project is that designer is free to design every single thing in it, from furniture to door hinges or monogrammed napkins and linen. They say it is beautifully eclectic but designer find it all to sing the same tune and that is what makes this home a very special one.

本案的设计师很幸运地可以自由设计项目中的每个细节，从家具到门铰链、字母花纹餐巾纸以及亚麻布料等。屋主觉得一切都很折衷，而设计师发现这里调性统一，这也是使之与众不同的原因。

SRI LANKA

BSL
BSL

ELABORATION CREATES CLASSIC

精心缔造经典

设计公司

DESIGN COMPANY

Interiorisimo

设计师

DESIGNER

Marisa Gallo

摄影师

PHOTOGRAPHER

Maxi Conesa

In this project,the design team recreated a contemporary ambient winking at the American 50s. A living room with different spaces related to leisure and social life surrounded by an atmosphere in which color, texture and the richness of materials are the project's concept. Copper regains its importance in the space through unique pieces. Different geometrical forms have been shaped into walls hangings achieving sculptural effects.

本案中，设计团队故意忽略美国50年代的风格，重新打造了一种当代氛围。客厅与休闲、社交生活的不同空间一起被这种氛围环绕，材料的色彩、质感与丰富是本案的设计理念。铜以其独特的形态恢复了在空间内的重要性。壁挂上不同的几何图形实现了雕刻般的效果。

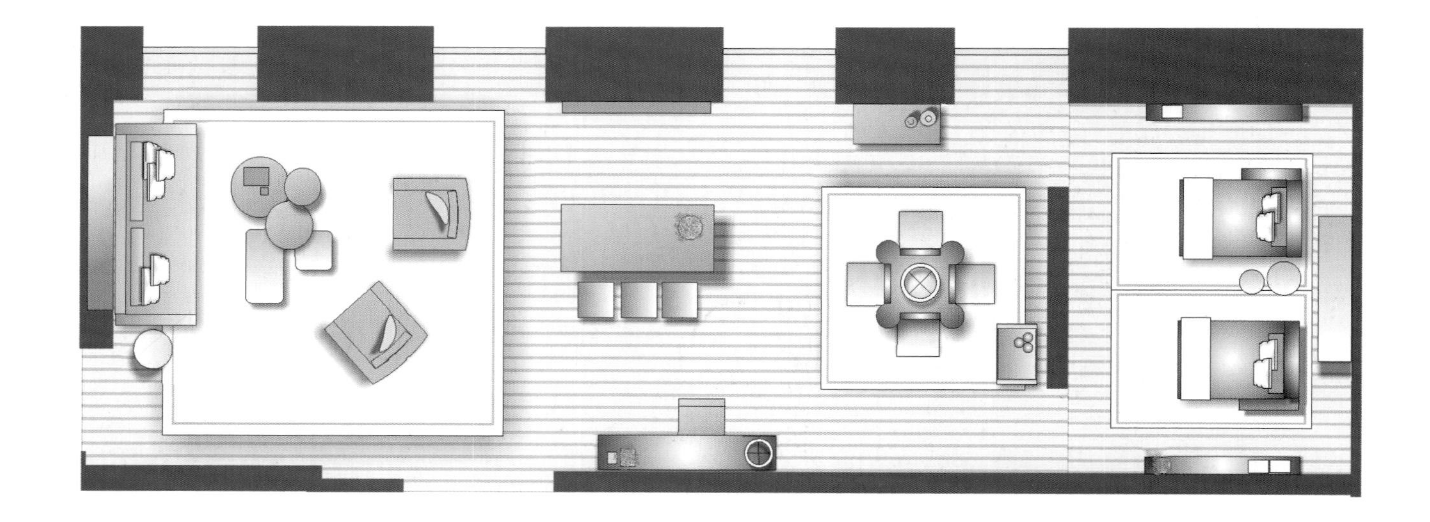

Wood is a characteristic element of this project, used both in the floor and furniture, that reflects warmth and sobriety. Another important material would be copper, featured in different applications, such as the incandescent lamp set falling over the bar. Walls are covered with wallpaper; it is of a neutral tone in the general room in order to project spaciousness, whereas, in the cinema area, a riskier darker shade has been chosen to provide it with depth and character.

木材是本案的典型元素，用于地板和家具上，反映出温馨庄重之感。另一个重要的材料是铜，应用在不同物件中，比如远眺着酒吧的白炽灯。墙壁上覆盖着墙纸，主要房间内运用中性色彩，使得本案宽敞大气，反之，在观影区则使用了暗色阴影，展现深度和特性。

SPAIN LIVING · COSTA DEL SOL

SUNSHINE AND CASTLE

阳光城堡

摄影师

PHOTOGRAPHER

Espacios y Luz Photography

This villa is being offered partly furnished and comes with ADSL, Satellite TV, alarm system, marble flooring, double glazing, hot & cold air conditioning throughout. There is a garage and private gated parking for at least two more cars.

这座别墅配有很多家具设备，比如宽带、卫星电视、报警系统、大理石地面、双层玻璃、冷热风空调等。这里还有一个车库，有两个以上的私人停车位。

You can have a chance to buy this unique immaculately presented south facing villa with 5 bedrooms which is on the sought after urbanization of Las Delicias in Coin. Only 5 minutes drive from la Trocha shopping center or 5 minutes in the other direction to the center of the charming Pueblo of Coin. 30 minutes drive to Puerto Banus, 25 minutes to Marbella, 30 minutes to Malaga airport.

This villa is unique because it was built to the present owners specification. The garden has been landscaped with easy maintenance in mind. There is a larger than normal swimming pool and plenty of sunbathing space as well as a secluded shady area. All the rooms are south facing which makes them lovely and bright.

这个独特而完美的朝南独栋别墅有5间卧室，在拉斯德里西亚斯的城市化之后被人们发现，现在有机会售出。距离购物中心只有5分钟车程，从另一个方向去迷人的普韦布洛市中心也只要5分钟。驱车30分钟可达巴努斯港口或马拉加机场，25分钟可至马贝拉市。

这栋别墅独一无二，因为它是按现任屋主的要求专属建造。花园造景简单，易打理维护。泳池比常规稍大，日光浴空间及阴凉区域很多。 所有房间都是南向，这令它们看起来明艳动人。

PEACE

On entering the property there is a well proportioned comfortable sitting room witch leads into a lovely bright kitchen diner both of these rooms lead out onto a large terrace that overlooks the swimming pool & garden. There is also a double bedroom on this level which also leads out onto the terrace.
Upstairs is the master bedroom suite with its own terrace. Downstairs are a further 3 bedrooms plus bathrooms and a second kitchen and sitting room. This could be used as separate self contained accommodation or kept as part of the house. In the garden there is a bbq area plus a work studio.

走进这栋别墅，匀称舒适的客厅映入眼帘，客厅通往明亮可爱的厨房和餐厅，这两处都通向可以俯瞰游泳池和花园的大型露台。露台旁还有两间卧室。

楼上是主卧套房及其阳台。楼下是三间卧室、卫生间改为厨房及客厅。这里可以独立使用，也可以作为整体房间的一部分。花园内有自助烧烤区域及一个工作室。

SPAIN LIVING · MALLORCA POLLENSA

CANRUO STARS

灿若星辰

设计公司
DESIGN COMPANY
Felip Polar. Estudio de Interiorismo

设计师
DESIGNER
Polar Felip

The house is situated in a privileged natural environment surrounded by a splendid Mediterranean and open to excellent views over the Serra de Tramontana. This unique location arises the need to generate an osmosis between adjoining elements and construction, such as stone, wood and light.

这个房子被地中海环绕， 有着优越的自然环境，面朝特拉蒙塔那山，可以远眺其优美风景。这一独特的地理位置需要各元素和建筑成分之间进行相互渗透，比如石头、木头和光线。

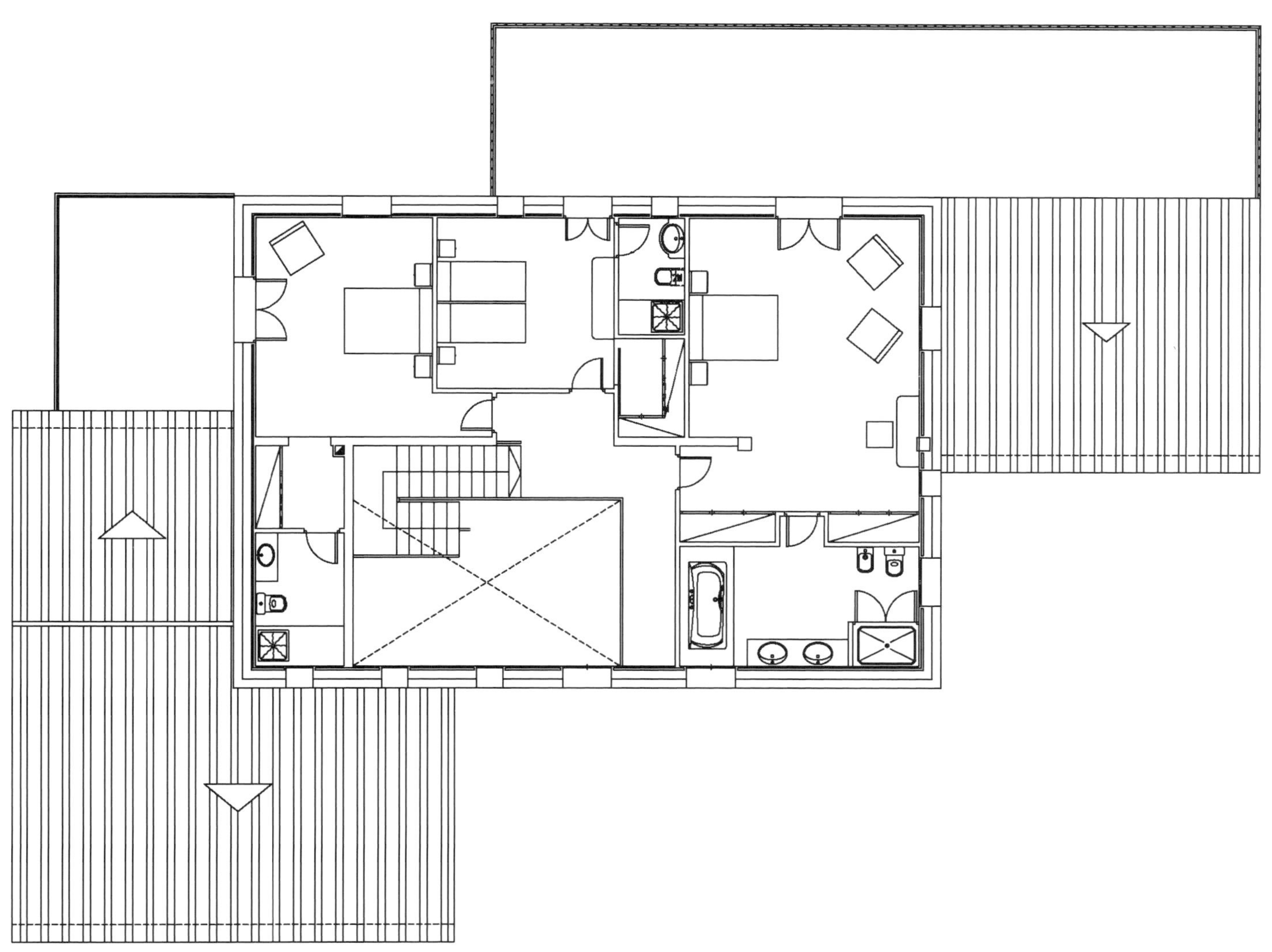

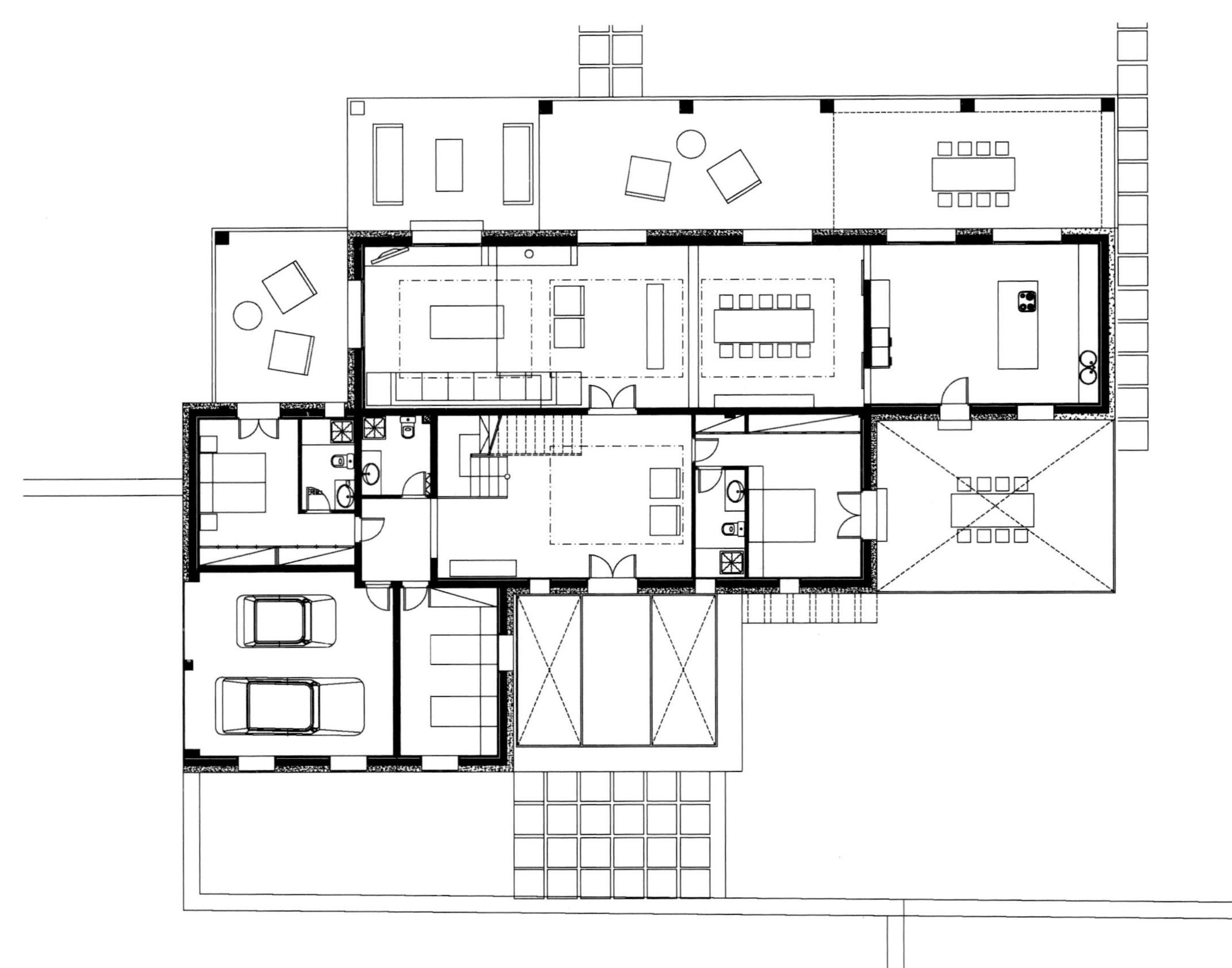

The transparent input of light in the house organizes the building into two rectangular volumes of geometric shapes. The first single plant that contains the sleeping area, rooms with their own bathrooms. The second volume, main and two heights, developed the living area downstairs and the master bedroom on the ground floor. The day area is located in one space on the ground floor, which contains the living room, dining room and kitchen, large windows open to the outside. The oak staircase have access to the main room.

The sleeping area develops ground floor into three bedrooms with bathroom, a guest toilet and cupboards. All rooms have access to an outdoor terrace.

This, together with the natural materials used for furniture, leading to a dialogue in harmony with the environment.

房子外面透进来的光使这栋建筑物的几何构造看起来像两个矩形空间。其中一个空间包括卧室和配有独立浴室的房间，另外一个空间有两层，都有起居区和主卧。白天活动的区域处于一个空间，位于第一层，它包括客厅、餐厅、厨房和向外敞开的大窗户。橡木楼梯可以通向主卧。

位于第一层的睡眠区包括配有浴室、公共卫生间和茶几的卧室。所有的房间都通向室外的露台。

以上设计和由天然原材料设计的家具共同展现了人与自然的和谐对话。

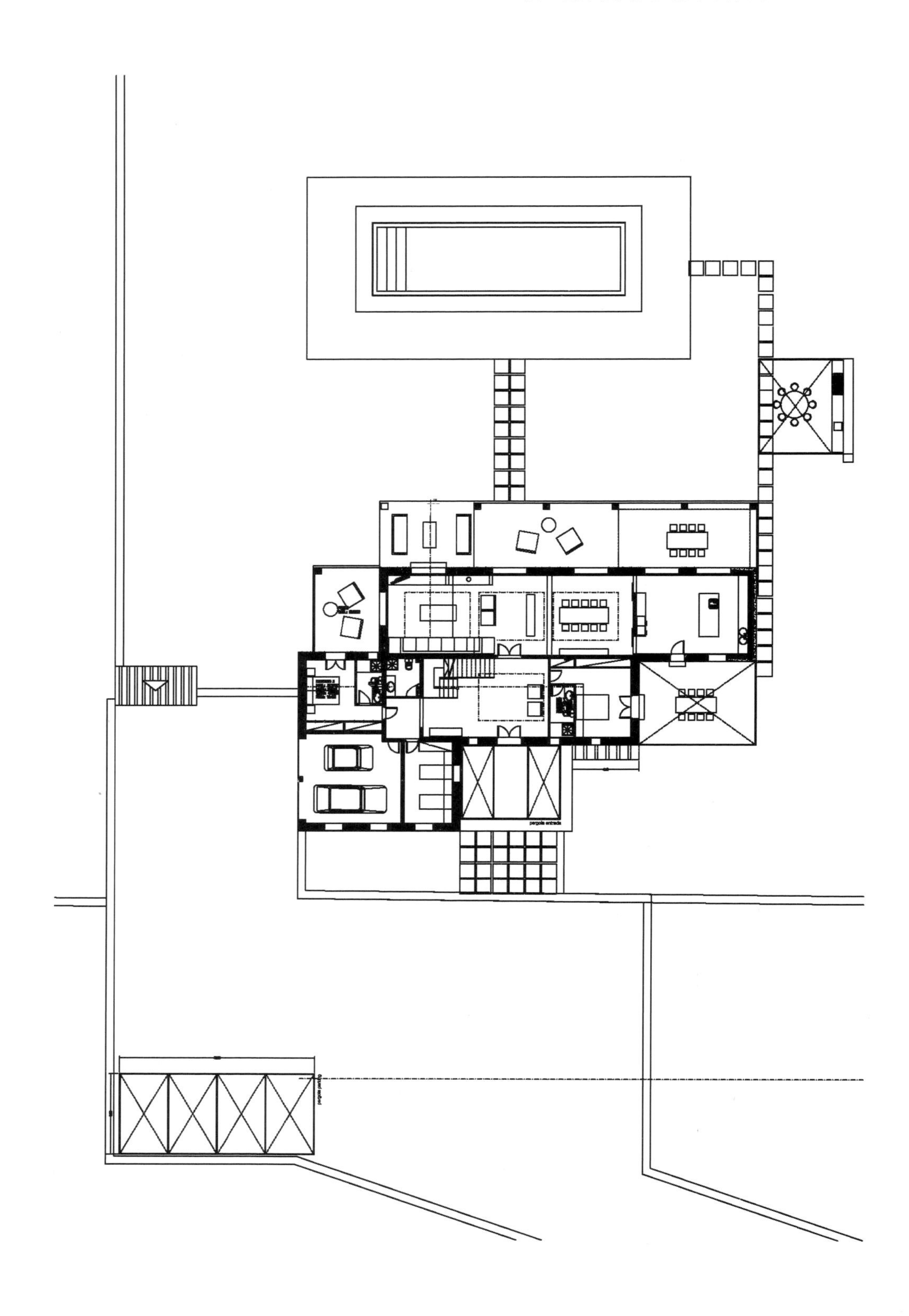

SPAIN LIVING · SEVILLE

TRAVELING IN A PICTORIAL WORLD

人在画中游

设计公司

DESIGN COMPANY

Amaro Sánchez de Moya S.L.

设计师

DESIGNER

Amaro Sánchez de Moya

摄影师

PHOTOGRAPHER

Martín García Pérez

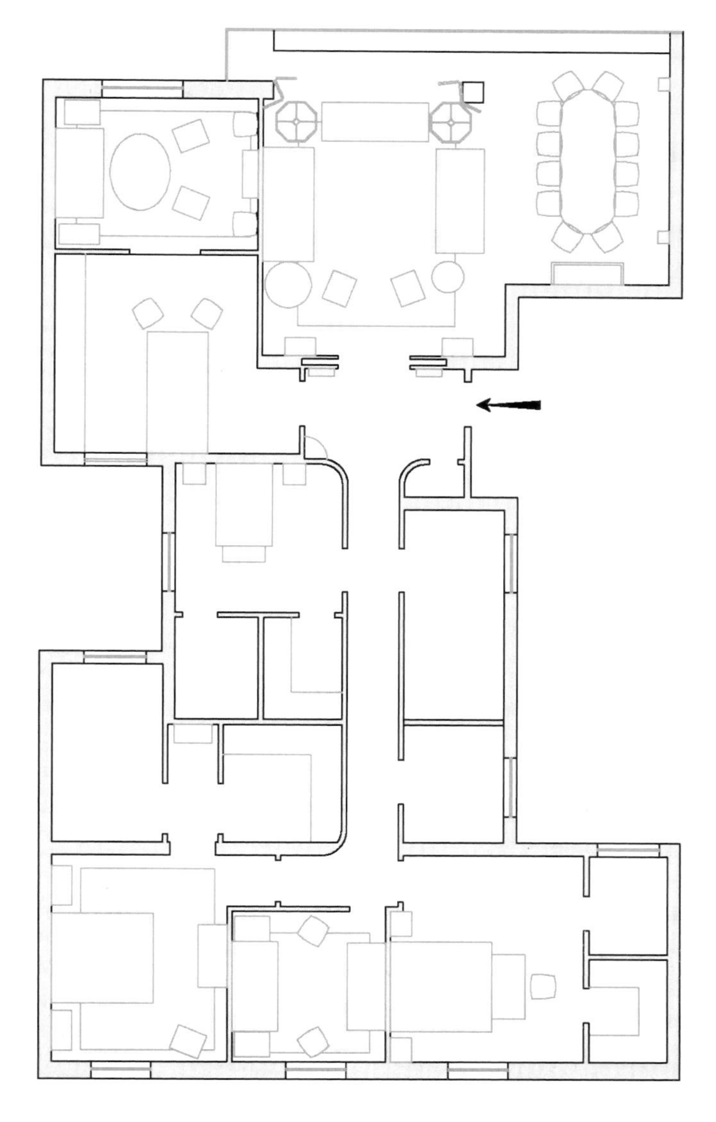

The idea for this family apartment house was to create a family atmosphere in the room had to a homogeneous collection of nineteenth-century painting of manners style. Spacious and comfortable, with colorful soft and cool spaces that serve as contrast to a collection that abounded tiny and small tables. Efforts were made through the choice of most modern fabrics and made furniture for the house, creating the ideal framework in which the painting collection could coexist naturally with needs specific to a family life without museum pretensions.

这间家庭公寓的理念是在拥有19世纪油画收藏的室内打造一种家庭氛围。宽敞、舒适、色彩缤纷且柔和清凉的空间与精致小桌上的收藏形成对比。选用现代布料并制成家具，打造理想的框架，这里的油画收藏可以与针对家庭生活的特殊需求自然共生，不加矫饰。

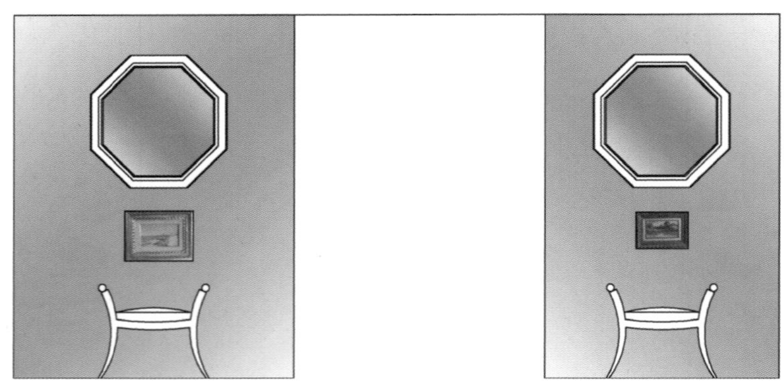

The house is set in different rooms suitable for every day of a family. The room was painted in bright aquamarine that reflects the warm sun of Sevilla and serves wide heel background painting collection, which had a very meticulous way in all walls of the house. The furniture, partly antique, partly specifically manufactured to design Amaro Sanchez de Moya. Carpets from the Royal Tapestry Factory cover almost all the floors of the house giving a great feeling of warmth.

不同房间的设置，满足了家庭的日常生活。室内漆涂鲜明的海蓝色，反射着塞维利亚温暖的阳光，并且为油画收藏提供背景，令房内所有墙壁熠熠生辉。部分家具是古董，另外一部分由设计公司定制而成。皇家壁毯厂的地毯几乎遍及所有楼层，给人一种非常温暖的感觉。

The kitchen and living room both are papered with a surprising role of blue vegetation. The private area of the house, bedrooms and office, are thought more soft and soothing colors, in which the pinks, grays and beiges predominate.

厨房和客厅采用了蓝色植被花纹壁纸，令人倍感惊艳。房内的私人空间、卧室及办公室运用了更多粉色、灰色以及米色为主的柔和色彩，起到抚慰人心的作用。

DELICATE GRADEN APARTMENT

雅致花园式公寓

摄影师

PHOTOGRAPHER

Espacios y Luz Photography

This is an apartment with direct exit to garden, and this is a popular style of residential complex in area, mainly for foreigners who like to stay living in Spain. Because people in Spain have a great weather mostly warm and not too cold or rainy as North of Europe, they have lot of blue sky days too.

Owner makes here a mix of Mediterranean with English style decoration. White walls help to give more luminosity to interior and bring a beam of natural light enter into the house. The dwellers need to just open the glass door to enjoy the big community garden, green and pool, in a quiet place.

本公寓直接通往花园，这种风格在住宅区内很流行，主要为喜欢居住在西班牙的外国人打造，他们愿意住在这里是因为这里气候温和宜人，不像北欧那样寒冷多雨，晴天比较常见。

屋主把这里打造成地中海风格及英式风格的混合风格。白色墙壁增添了室内的明亮度，让自然光线可以照射进来。居住者只需要打开玻璃门，就可以欣赏大型社区花园美景、绿地与泳池共处一片宁静祥和之中。

Kitchen is not big because foreigners mostly eat out at restaurants with healthy Mediterranean food, fish, vegetable, and local products. As sometimes owners are living for long time in their countries , they usually have grills in windows and doors.

因为外国人大多选择外面的餐厅食用地中海健康食品，如鱼、蔬菜和当地特产，所以厨房的面积并不大。有时，屋主在自己的国家生活久了，就习惯在门窗上设置烧烤架。

This property is placed in Calahonda area which is a green area embracing the sea and beach, with a precious climate. Close to the luxury Marbella city and 20 minutes of Malaga city driving a car. The apartment has 2 beautiful bedrooms and 2 bathrooms. It is situated in the lower western border of Calahonda amidst tranquil settings. Within 10 minutes walk of the beach, shops, bars, restaurants and all facilities making Jardines de Calahonda an ideal holiday destination for families, couples and groups.

Jardines has its own local neighborhood on-site bar, handy for enjoying a refreshing drink, a light snack or an ice cream. This two-bedroom ground floor apartment is beautifully furnished to an exceptionally high standard. Fully refurbished bathrooms (with showers only, no bath) and kitchen, it has air conditioning / heating in the lounge and ceiling fans in both bedrooms.

这套公寓位于西班牙的绿地区域，环抱海洋和沙滩，气候宜人。毗邻奢华的马贝拉市，距马拉加市也仅有20分钟车程。

该公寓包含2间迷人的卧室及2间浴室。坐落在下游的西部边界，安静平和。步行至沙滩、商店、酒吧、餐厅及其他地方仅需10分钟，使得这里成为理想的家庭、情侣、团体度假圣地。

这里有自己本地的现场酒吧，方便享用消暑饮品、小点心及冰淇淋。这座有两间卧室的公寓装修华丽，标准异常之高。浴室（只有淋浴，无浴缸）和厨房全部翻新，休息室内设冷暖空调，两间卧室都安装了吊扇。

The lovely sunny west facing private terrace has direct access to the beautiful kept landscaped gardens and communal pool. The property also offers WIFI and internet enabled British TV. The perfect choice for all age groups.

西向私人阳台可爱迷人、光照充足，可直接通往美丽的花园及公共泳池。这里还提供无线网络及英国网络电视，符合所有年龄段的人的选择。

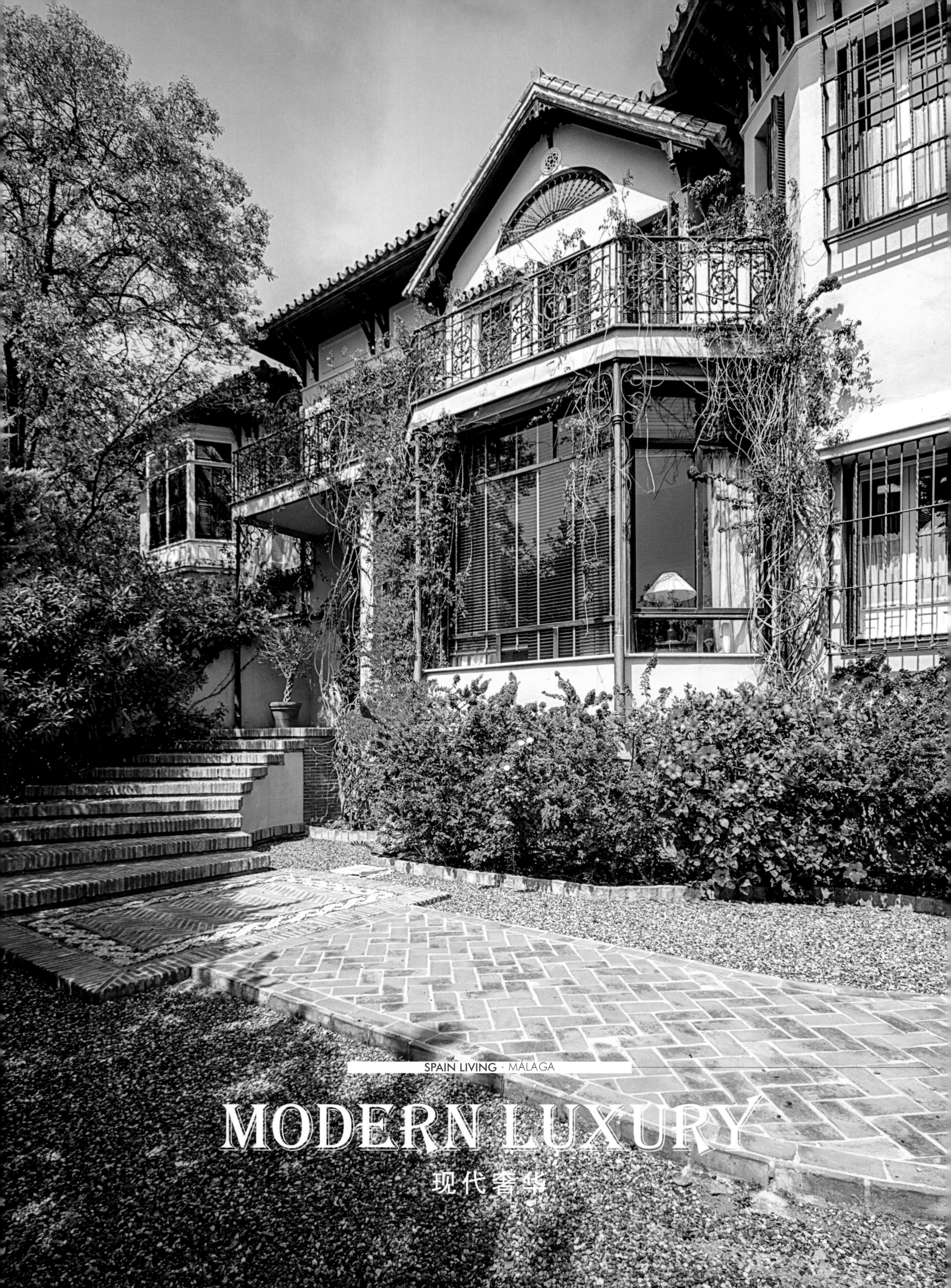

SPAIN LIVING · MÁLAGA

MODERN LUXURY

现代奢华

摄影师

PHOTOGRAPHER

Espacios y Luz Photography

Monte Miramar is a 1500 square meters plot and 450 square meters build house in 2 levels, 3 bedroom plus service bedroom. This building is a historic building and is government protected because it has 100 years old and possibly built by one of famous architects of earlier 1900 who did the prime streets and commercial buildings in Málaga Historic and commercial Town Mr. Fernando Guerrero Strachan 1879-1930. So owners need to keep the facade and structure as origins.

这个项目规划用地1500平方米，建设用地450平方米，共两层，包括三间卧室，两间服务室。该建筑具有历史意义且受政府保护，因为它100岁高龄，可能是20世纪早期著名建筑家费尔南多·格雷罗·斯特罗恩的作品，他曾设计了马拉加极具历史性与商业性的城镇中的主要街道和商业建筑。所以，屋主必须保留其原始外观与建筑结构。

Located in a hill with nice views to Malaga's Bay and sea, city and port, one of the best places and rich places of Málaga City, close of features. Monte Miramar has a big garden with 2 fountains and different king of forest and flowers, chipmunks run on top of cups of trees. Interior is decorated with a mix of different furniture of different cultures, countries and styles richly mixed, we can see a nice Asian umbrella beside French style chairs. Elegance is living in each corner of the house and warm colors combined with ancient woods of furniture and ceilings.

它位于一座小山上，可以欣赏马拉加的海湾、海域、城市和港口，风光旖旎，是马拉加市数一数二的富饶之地。蒙特·米拉玛拥有一个大型花园，两眼温泉、不同种类的植被花卉，还有穿梭在树干上的花栗鼠。室内装饰着不同文化、不同国家、不同风格的各类家具，这些家具在屋内混合搭配摆放着，比如说，法式椅子旁有一把精致的亚洲雨伞。优雅遍布于屋内，暖色调与家具、屋顶上古老的木材完美结合。

TRANQUILITY

静谧

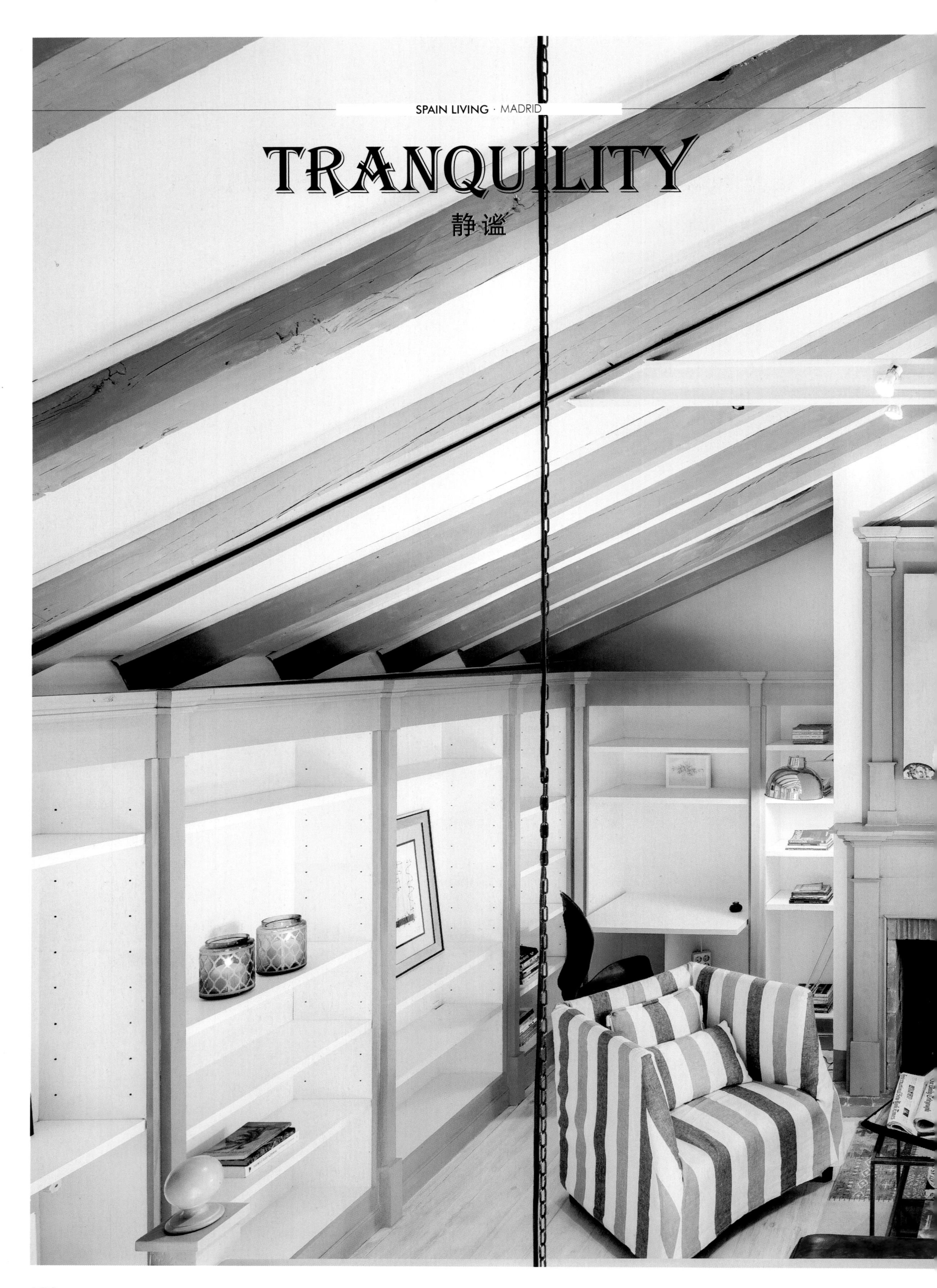

设计公司

DESIGN COMPANY

THE SIBARIST PROPERTY & HOMES

设计师

DESIGNERS

Sylvia and Penélope Girón

摄影师

PHOTOGRAPHER

Pedro de Agustín

This elegant penthouse has been conceived as a good place for tourists to have a different experience. Located in the center of Madrid, it is the first brand of The Sibarist Property & Homes and is flagship brand.

After a series of interior designs in August 2014, the attic of 140 meters is divided into three floors and has a capacity of 7 people. It has 3 bedrooms, 2 bathrooms, and 2 living rooms with fireplaces. Its ceilings are high and the entire structure is wooden beams. From its main hall, you can see the dome of the Teatro Real in Madrid. All rooms have windows and pouring light, with absolute tranquility.

The house has been carefully decorated by the interior team from The Sibarist, specializing in furniture design and interior spaces. They have taken into account even the smallest detail. The interior design is created by Sylvia and Penelope Giron. Sylvia is the founding partner of The Sibarist and Penelope owns Casa Nativa.

这个雅致的阁楼位于马德里的中心，一直都被认为是能为游客提供不同体验的好住处，是Sibarist房产公司的第一品牌，也是它的旗舰品牌。

经过2014年8月份一系列的室内设计后，这个140平方米的阁楼被分为了三层，可以居住7个人，有三个卧室，两个浴室，两个带有壁炉的客厅。天花板很高，由木梁组成整个建筑结构。站在走廊上，你可以看到马德里皇家剧院的圆顶。所有的房间都有窗户，光线射入，极其宁静。

这个房子是由Sibarist房产公司的设计团队精心设计的，他们专门从事家具设计和室内设计，即使是最小的细节问题，他们也会考虑在内。西尔维娅和佩内洛普·吉龙共同完成了室内设计，西尔维娅是Sibarist房产公司的创始合伙人，佩内洛普创建Casa Nativa。

ARROYO
BROTO
CANOGAR
CHIRINO
GENOVES
GORDILLO
IGLESIAS
LOPEZ
PLENSA
VALDES

SPAIN LIVING · BARCELONA

CLASSIC BUT GLAMOROUS

迷人的经典

设计公司
DESIGN COMPANY
Ana Ros

设计师
DESIGNER
Ana Ros

摄影师
PHOTOGRAPHER
Nicolás fotografia

It is a noble floor of the extension of Barcelona, it is a classic floor moldings and has 300 meters.

It has three suites complete with bathroom and dressing, large hall and two large rooms and a terrace. The decor is classic but glamorous, giving it a modern touch. The decor is based on the French Empire era and also in a living room a vintage feel.

It is timeless to give all the decoration, playing with old classic furniture which make it original and unique. The designer has also tried through wallpaper, carpets and mirrors to give a very warm and cozy feel.

这间贵族公寓位于巴塞罗那，造型经典，面积300平方米。

三间套房带有独立卫浴及更衣室，另有一个大型门厅、两个大房间和一个露台。装修风格经典而不失魅力，拥有现代化的感觉。该风格以法国君主制时期为背景，客厅中便是复古氛围。

装修永恒不过时，古老的经典家具令这里看上去更加与众不同。设计师还尝试通过壁纸、地毯和镜子等打造出温馨之感。

CASTLE'S MEMORY

古堡记忆

设计公司
DESIGN COMPANY
Renova Constructores

设计师
DESIGNER
Jörg Schmitz

摄影师
PHOTOGRAPHER
Oksana Krichman

Built in Art Nouveau style, designed by the architect Gaspar Reynés i Coll (with Jaume Alenyar), the property was listed as being of historical interest in 2005 by the Regional Government of Mallorca (Consell Insular). The architect has also designed the well known building Hostal Cuba in Santa Catalina, also built in 1904, and also listed as a building of historical and cultural interest.

该建筑定位为新艺术风格，由建筑师加斯帕·瑞奈斯·科尔与詹姆·阿恩雅共同打造，被马略卡岛地方政府列为2005年度历史古迹。建筑师加斯帕·瑞奈斯·科尔还在圣卡塔利娜岛设计过知名的Hostal Cuba大楼，同样建于1904年，都被列为历史文化古迹。

1904

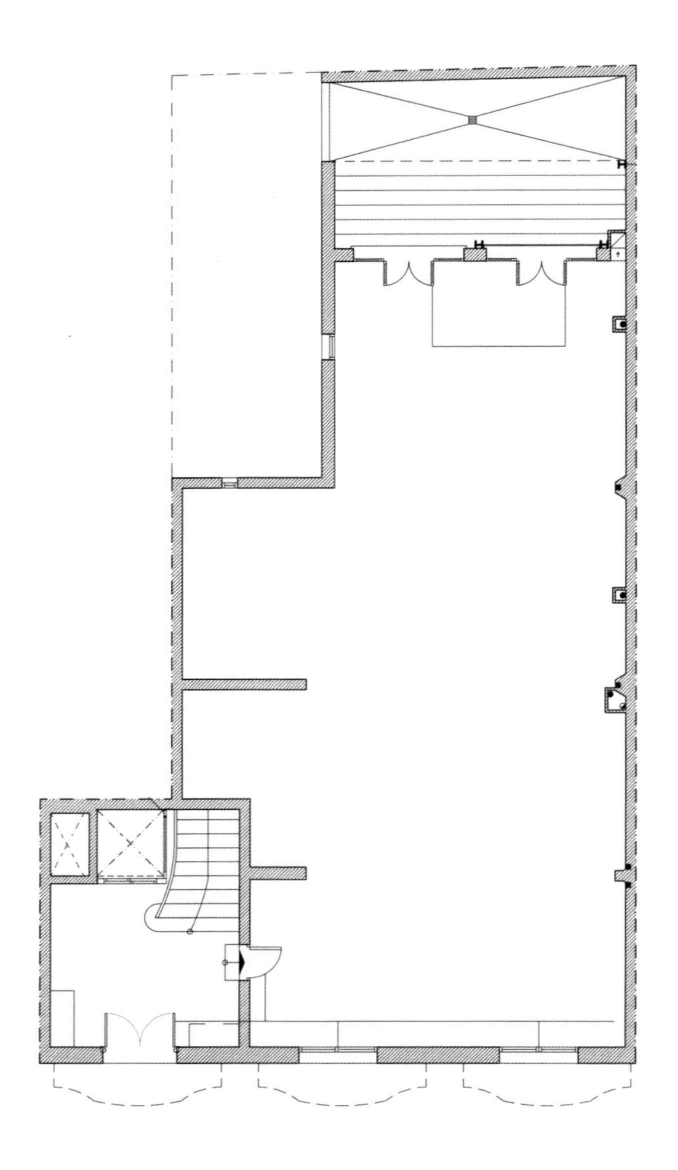

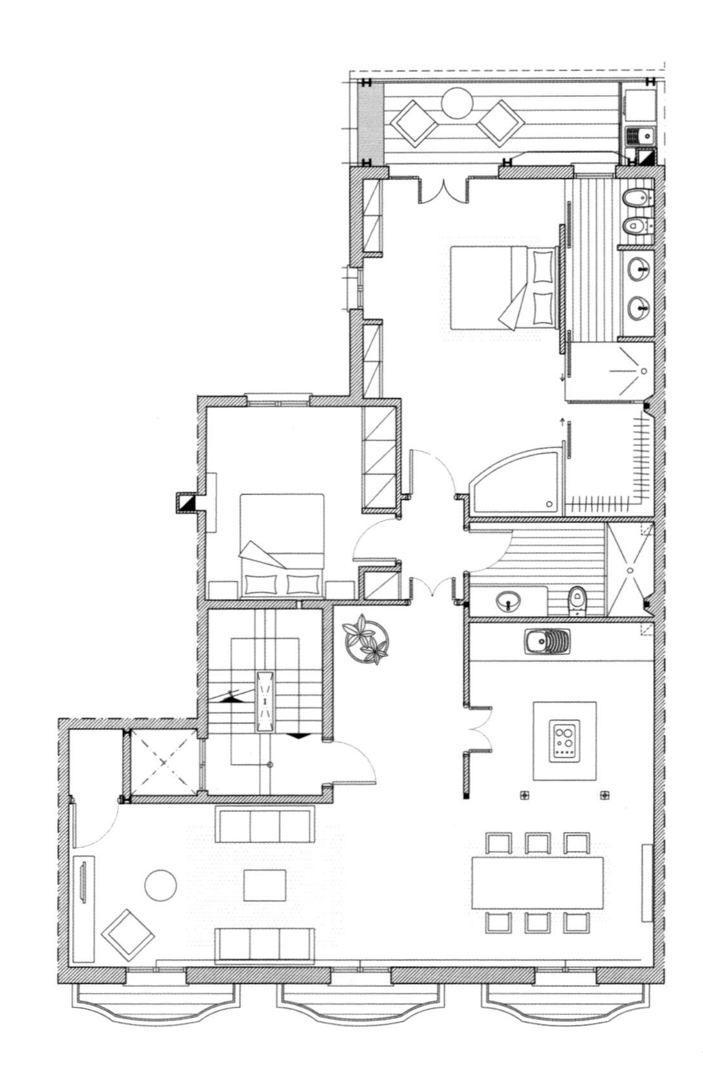

This project displays many of these typical Art Nouveau signs. Visible in its facade of subtle beauty, built in Majorcan Mares sandstone, it shows intricate floral ornaments around the windows, under the balconies and at the impressive top of the building as well as a decorative floral sign with the building year 1904 carved in. Also the balcony railings, being the only tri-dimensional element of the facade, are typical for this architectural style. Inside the house the most impressive features are the beautifully ornamented pillars that naturally divide the living space.

Lacking solid foundations and taking into account that the building was tilting towards one side, designer had to secure it fully in order to carry out certain phases to make an adequate foundation, at the same time making a parallel structural backing. With a master bulk wall system strengthened by wooden rafters we had to demolish the 72 square meters of one of the four bulk walls with the ensuing support of the building by the rest of the walls and provisional metal supports.

The protected elements were adapted to a new architectural concept so as to preserve them without disavowing current modernity.

The building suffered from various pathologies in its structure that we had to correct including wooden joists that were in advanced stages of decomposition, cracks in master walls and iron pillars that were very deteriorated.

本案陈列着众多典型的新艺术标志。由马略卡岛的马雷什砂岩打造的立面具有精妙之美，在窗边、阳台下以及建筑顶部展现着错综复杂的花饰，同样还有一个1904年随楼而刻的装饰性印花。阳台栏杆是立面唯一的三维元素，也是这种建筑风格的典型特征。屋内给人留下最深刻印象的便是自然划分生活空间的精美装饰柱。

这栋建筑缺乏实体基础，考虑到其向一面倾斜，设计师不得不将其全部固定，以采取措施打造充足的基础，同时建造一个平行结构的背衬。因为木制椽强化的主体墙系统，设计师必须拆除72平方米的四面大体量墙壁之一，并运用其余墙壁及临时金属支架为该建筑提供安全支持。

受保护的元素适用于一种新建筑理念，在不否定当今现代化的同时保存其自身。

该建筑在结构上遭逢一系列病变，设计师必须进行更正，包括处于腐烂变质晚期的木制托梁、主墙体上的裂缝以及蚀薄的铁柱。

The three apartments were accommodated with maximum comfort such as air conditioning via conducts, radiator heating, a water deposit installation with pump so as to assure plenty of supply at all times. What stands out the most within the work carried out is the restoration of the facade, the wood carpentry and the interior decoration, the complete removal of the rafters and their new placement, substituting the sandstone (mixture of sand and chalk) slabs with viroc-plaques (cement-wood) in order to lighten the joists, and the manual recovery of more than 200m^2 of hydraulic tiles that were more than one-hundred years old and later were put back by orders from the Heritage Trust.

三间公寓的舒适度方面均达到最大化，比如空调、暖气供暖、蓄水泵，以确保始终拥有足量供给。最引人注目的是立面的恢复工作，木工、室内装饰、椽木的完整去除以及重新定位，取代了砂岩板，以便提亮托梁，并且手工复原了200余平方米的液压瓷砖，那曾有百余年的历史，后被遗产信托重置原位。

SIMPLICITY

简约质朴

设计公司
DESIGN COMPANY
Belen Ferrandiz Interior Design

设计师
DESIGNER
Belen Ferrandiz

摄影师
PHOTOGRAPHER
Montse Garriga

This is a ground floor flat with a lovely garden in one of the most exclusive areas in the outskirts of Madrid. The owner is a lovely lady with a very big family, whose first premise was to have a house to entertain all her family. In the concept of this house, the most important issue was to create a home to be enjoyed and shared by the family. This was the main reason for the designer to focus on the wide common areas, to design a lovely and beautiful home.

屋主是一个和善的女士，有一个大家庭，她的首要前提是这个房子能招待她所有的家人。这个一层平房有个可爱的花园，位于马德里郊区的黄金地区。这个房子最重要的设计理念是创造一个全家人共享的环境。这也是设计师为了设计一个可爱而美丽的家而专注于这块公共区域的原因。

This house did not need a renewal, it was newly constructed. And the owner did not want to get in masonry works, things the designer projected were to separate the living room areas with huge sliding doors to create one big area to live in and share, and also separate the dinning room with the kitchen by a glass . All areas, including dinning room, living room and even the porch, needed to host the largest number of people, this is the reason why there are large get-together areas and impressive dinning tables.

由于房子是新建的，所以不需要翻新。业主也不想有过多的砌筑工程，设计师所要做的就是用大滑动门隔开客厅，用玻璃隔开餐厅和厨房，预留更多的独立空间给这个大家庭。所有的区域，包括餐厅、客厅，甚至玄关，都需要能容纳大量的客人，这也是这个家里有多个聚会场所和令人印象深刻的餐桌的原因了。

TRADITIONAL MEDITERRANEAN STYLE

传统地中海风

设计公司

DESIGN COMPANY

MELIAN RANDOLPH S.L.

设计师

DESIGNERS

Victoria and Sylvia Melian Randolph

摄影师

PHOTOGRAPHER

Martin Garcia Perez

The house is located in a resort called Sotogrande which is in the South of Spain.

The original house was built in 2000 for a family with young children, by architect Joaquin Mier Enriquez. It had a Main House and Porch with a Garden and Pool Area. We were called in a few years later to add a Pool changing room with sauna, an outdoor glassed dinning room, a study, a Guest House and finally to re-design the kitchen area.

The house itself has very Andalusian architecture, with terracotta floors, whitewashed walls and tiled roofs. We wanted the new sections of the house to blend into the existing so they would look like a unit rather than additions. The windows and doors are wood, but the new dining area has an iron and glass closing and a new porch covered in bamboo for shade.

这栋房子位于西班牙南部一个名为索托格兰德的度假圣地。

始建于2000年，这个房子由建筑师杰奎因・米尔・安立奎为一个有小孩子的家庭而造，包括房子主体、门廊、花园及泳池。几年后，屋主找来设计师增建一间带桑拿的泳池更衣室、一间室外玻璃餐厅、一间书房、一间客房并重新设计厨房。

房子本身是具有浓烈安达卢西亚风格的建筑，陶瓦地面、石灰墙壁以及瓦屋顶。设计师希望新加入房子的部分可以与现有部分和谐共存，比起附加物来说，更像是一个整体。门窗都是木制，而新餐厅区域由铁与玻璃密封，新门廊覆竹子以庇荫。

LARGE INFORMAL LIVING ROOM (WHITE WALLS)

大型非正式客厅（白色墙壁）

This room is used by the family and children's friends to watch TV, play games and listen to music.
The sofas come for Ascensión Latorre and we had cotton white covers which can be cleaned easily but also give a summery look.
The coffee table comes from a flea shop in Madrid.
The armchairs are a Paulistano design and are leather and metal.
The carpets come from Morocco. They are hand weaved in wool.
The painting is by Ferrán García Sevilla who is a Spanish Artist and the painting is from the 80's.

这个房间被家人和孩子们的朋友用来看电视、玩游戏和听音乐。

沙发来自Ascensión Latorre，设计师覆以白色棉织物，易于清洁且能带给人夏日般的感受。咖啡桌来自马德里一家跳蚤商店。

皮革加金属合金扶手椅是Paulistano的设计。

手工编织羊毛地毯来自摩洛哥。

油画出自西班牙画家Ferrán García Sevilla之手，作于80年代。

OUTDOOR DINNING ROOM
室外餐厅

Old Vintage dinning table with wooden legs and metal top. Chairs in metal from La Europea antique shop in Madrid.

复古餐桌拥有木制桌脚及金属桌面。餐椅由金属打造，来自马德里La Europea古董店。

PORCH
门廊

Rattan sofas, ironwork chairs and stools and wood table form Casa y Campo in Madrid. The carpets are Moroccan hand woven wool from Lola Coronel.

藤制沙发、铁艺座椅与长凳、木桌均来自马德里的Casay Campo。摩洛哥手工羊毛地毯来自Lola Corone。

SWIMMING POOL
游泳池

The pool was done with small green tiles on the inside in order to give it a "pond color" look instead of turquoise. The three sun beds were made by J.J. Cuenca in Málaga. The earthen ware hand made pots and the gardens were designed by garden designer Jesús Gómez from Jerez (Spain)

泳池内部采用小型绿色瓷砖，使其显示出"池塘的颜色"，而不是蓝绿色。三张日晒床由J.J. Cuenca在马拉加制造。手工陶器及园林景观均由西班牙赫雷斯市的景观设计师Jes ú s G ó mez设计。

ENTRANCE HALL CORRIDOR
入口大厅门廊

Beautiful and colorful art piece by Ian Gillick. The carpet is Moroccan from Lola Coronel and the chair is Portuguese wicker from Làpety.

这里有Ian Gillick的艺术品、来自Lola Coronel的摩洛哥地毯以及L à pety的葡萄牙柳编椅。

INDOOR DINING ROOM
室内餐厅

This dining room is used mainly in the winter months. The table and chairs are XVIII French and were bought from Los Gusano which is an excellent antique shop in Madrid. The carpet is Moroccan hand woven wool from Lola Coronel. The industrial lights hanging from the celling are vintage and come from La Europea shop in Madrid.

这间餐厅主要用于冬季。餐桌椅是法国18世纪风格，是从马德里一家出色的古董店买来的。摩洛哥手工编织羊毛地毯来自Lola Coronel。天花垂吊的工业风复古灯具来自马德里的La Europea商店。

KITCHEN
厨房

The architect for this renovated and enlarged kitchen is Eduardo Dorissa. The floor are stone and the hand made tiles on the walls come from Lisbon (Portugal).
The table was made to measure for the Melián Randolph studio. The sofa is covered in Joseph Frank linen fabric. The cane chairs come from Juan José Cuenca in Málaga. The green ones are a design by Hans Wegner and the white ones are old French. The side tables in wicker come from J.J. Cuenca also.
All the ceramic plates are old Spanish hand painted and were bought at the auction house Subastas Goya from Madrid.

负责翻修并扩建厨房的设计师是Eduardo Dorissa。地面由石材制成，墙上的手工瓷砖来自葡萄牙里斯本。

桌子由Melián Randolph工作室定制。沙发上覆盖着Joseph Frank亚麻织物。藤椅来自马拉加的Juan José Cuenca。绿色的椅子由Hans Wegner设计，白色的椅子来自古老的法国。柳条边桌同样出自J.J. Cuenca。

所有陶瓷板都是手绘古西班牙风格，购于马德里Subastas Goya拍卖行。

WHITE BEDROOM
白色卧室

The floors are wood and the walls were painted with local whitewash paint with grey pigment added to it. The beds were designed by Melián Randolph and were manufactured in Seville by Chisel and Rose. The collection of vintage mirrors come from Vintage 4P in Madrid. The cow hide carpet comes from Taxidermia El Ciervo in Gerona (Spain).

木地板搭配当地白灰粉刷的墙壁，墙壁上还有灰色颜料。床由Melián Randolph设计并由Chisel和Rose在塞维利亚制造。复古镜子的收藏来自马德里Vintage 4P。牛皮地毯来自西班牙赫罗那市的Taxidermia El Ciervo。

PINK AND GREEN BEDROOM
粉色及绿色卧室

The headboards are covered in Manuel Canovas fabric from Gastón y Daniela in Madrid.
The carpet is hand woven Moroccan wool from Lola Coronel. The bed covers in Indian cotton are from Aunty B in Madrid.

床头板附着物采用马德里Gastón y Daniela的Manuel Canovas面料。

手工编织摩洛哥羊毛地毯由Lola Coronel制造。印度棉床单床罩来自马德里Aunty B。

The new bathrooms have cement colored tiles on the floors which were used at the beginning of the century and add a touch of color to contrast with the terracotta elsewhere, and the new kitchen has Portuguese glazed tiles on the walls, again to add color into the house.
Everything else was respected and is very Mediterranean and traditional.

新浴室的地面都采用本世纪初流行的水泥彩色瓷砖，这些色彩用来与各处的陶瓦形成对比，新厨房的墙壁上贴着葡萄牙琉璃瓦，又为房间增添了色彩。

其余一切均得以保留，传统地中海风味非常浓郁。

BATHROOM WITH BATHTUB
浴缸浴室

The tiles are painted cement hand made and come from Alvaro Guadaño in Madrid as well as the sinks and bathtub. The interior designers put a wooden skirting all around the walls. The mirrors are Laura Ashley.

瓷砖由漆面水泥手工打造，与水槽、浴缸一样来自马德里Alvaro Guadano。室内设计师在墙壁周围放置了木制壁脚板。镜子出自Laura Ashley。

BATHROOM WITH ROUND WASHBASINS
圆洗手池浴室

The floors are grey wood. The piece of furniture specially designed to fit the wash basins is French and comes from Provence & Fils. The mirrors were set into the wall and are metal, vintage and come from a boat. The decorators found them in a flea market.

这里使用了灰色木地板。为与洗手池匹配而特别设计的法式家具来自Provence & Fils。嵌入墙壁的复古镜面带有金属，源自一艘船。这些都是装潢师在跳蚤市场买来的。

RUSTIC FAMILY RESIDENCE

淳朴家居

设计师
DESIGNER
Dafne Vijande

摄影师
PHOTOGRAPHER
Santiago Moreno

HISTORIA
ECONOMICA Y SOCIAL
DEL MUNDO
4
1840 1914

This house is designed for a large family to spend weekends there. It is located on a large estate in the middle of Andalusia. The architecture of the house is based on the ancient Andalusian farmhouses. However, the interior is more current and modern with large windows to enjoy the view of the countryside, especially in winter that days are colder. The house has several wings or areas: the main, are the common areas and the master bedroom and the wings of the house is where the service area and the children's area. Orientated to the southwest are plants and vines that climb up the facade. The house is ecologic, it has solar panels that give light, gas tank for radiant heat (on the floor) and well for water.

本案是为一个大家庭设计的，以便于他们可以在这度过周末。它坐落于安达卢西亚区中心位置的一片大住宅区内。它的外观是根据古安达卢西亚区农舍而建的。但是里面的设计很现代化，透过大窗户可以欣赏乡村美景，尤其是在寒冷的冬天。这个房子可分为三片区域，其中主要区域包括公共区域和卧室，房子两侧是服务区和儿童区。在西南面，人们可以看到一些植物和藤蔓爬满了围墙。本案的设计很环保，可以用太阳能发电，一楼用煤气作为燃料，用井取水。

ARROZ

The fireplace is constructed with a series of stones such as Roman foundations. All building elements were made with materials and local craftsmen. The flooring is designed with marble, equipped with radiant heating.

The living room has high ceilings and has a large fireplace that makes heating in that space. The living room divides the house into 2 areas: the common areas and the bedroom area. All public areas have large sliding doors to join or separate the spaces, according to the needs in each moment. So the living room is attached to the dining room and the kitchen by a sliding door. On the other side of the living room, there is a small living room which is attached to the bedroom by sliding doors. The furniture is inherited from ancient.

壁炉的设计运用了一系列的石材，比如罗马地基。所有的建筑材料都是用原材料加工，请的当地工匠。地板使用大理石建造，房子配有供暖设备。

客厅的天花板很高，有一个很大的壁炉可供加热取暖。客厅将房子分为两个区域，即公共区域和卧室。所有的公共区域都配有滑动门，根据不同需要可将空间聚拢或分隔。滑动门将餐厅、厨房与客厅连接在一起。在客厅旁边有一个小客厅，通过滑动门与卧室相连。家具的设计沿袭了传统风格。

Above the lounge, there is a loft with a library where you can see the views of the whole valley of the river Guadalquivir. The bedrooms are spacious, using antique furniture inherited and others designed specifically for the home.

在休息室上面有一个配有图书馆的阁楼，在此处可以远眺瓜达尔基维尔河的整个峡谷。卧室很宽敞，选用了定制的传统家具。

SUMMER HOUSE
夏舍

设计公司
DESIGN COMPANY
Belen Ferrandiz Interior Design

设计师
DESIGNER
Belen Ferrandiz

摄影师
PHOTOGRAPHER
Montse Garriga

This is a chalet house in one of the most exclusive areas in the North outskirts of Madrid. The owner is a couple with three children. They focused on creating a young house to enjoy the weekends and summertime. So it needed to be clear, fresh and open. The couple was involved in all the works and details of the project with a lot of enthusiasm.

In the concept of this house, the most important issue was to create a house easy to live, easy to share with friends and easy to live with children. This was the main reason why the designer decided to demolish all the walls in the ground floor. The most important thing to highlight in this project is the influx of light. This is what the designer got from the demolition of all the walls.

这个小木屋位于马德里北部郊区的黄金地区。屋主是育有三个孩子的一对夫妻。他们专注于设计一个能享受周末和夏日时光的小屋。因此，这个房子必须光线好，空气清新，视野开阔。值得一提的是，这对夫妻非常热衷于这个项目所有的工作和细节。

这个房子的设计理念中，最重要的问题是要创造一个方便生活和接待朋友，容易与孩子相处的环境。这也是设计师决定拆除一楼所有墙壁的原因。这个项目最吸引人的是大量光线的照射，而这得益于设计师拆除了所有的墙壁。

SPAIN LIVING · BARCELONA

A LEISURE HOME

休闲之家

设计公司

DESIGN COMPANY

Meritxell Ribé - The Room Studio

设计师

DESIGNER

Meritxell Ribé-The Room Studio

摄影师

PHOTOGRAPHER

Maurici Fuertes

Complete reform of this house, keeping the original elements itself: cornices, vaulted ceilings, floors, fireplaces, doors, etc. returning all the glow of the property that was in the 50s, but incorporating new technologies such as home automation, cinema and music.

This house is designed in a leisure style, reflecting the easygoing character of its owner. There is no excessive decoration, but one may have a comfortable feeling while touring the house.

这间公寓的设计保留了它原本的要素：飞檐、拱形天花板、地板、壁炉、门等。这些元素使它重现了19世纪50年代的辉煌。但与此同时，它的设计也结合了现代新科技，比如家庭自动化科技和影音技术的运用。

本案的设计为休闲风，反映出屋主亲和的个性。并不需要过分的装饰，但倘若置身其中，便会有温馨舒适的感觉。

TRAVEL AROUND THE WORLD

环游世界

设计公司

DESIGN COMPANY

MASFOTOGENICA INTERIORISMO
www.masfotogenica.com

设计师

DESIGNERS

PILI MOLINA NAVARRO &
MYRAMAR

摄影师

PHOTOGRAPHER

CARLOS YAGÜE RIVERA for
MICASA (Hearst Magazines Spain)

Located in Benalmádena, Málaga, Spain, this show flat was asked to create a space that looked lively and naturally. Constructor and Developer MYRAMAR(http://www.myramar.com/) wanted an eclectic furnishings decoration created with pieces from many different sources to attract young families, national or foreign.

Located on the Costa del Sol where people from many countries visited, it is perfect to design a house ready not only for vacation, but also for people who decides to live all year long in this area. So this apartment is in the "Travel around the world" style.

位于西班牙马拉加，这个样板房意在打造一个活泼自然的空间。MYRAMAR设计团队想使用折衷主义的家居装饰来吸引国内外的年轻家庭。

这个房子位于太阳海岸，世界各地的人们都来这旅游过。所以在这里设计一个房子，不管是为人们度假提供场地，还是为决定长期居住在此的年轻人提供居所，都是再好不过的了。因此，这个公寓就有了“环游世界”的风格。

E
C B
D L N
P T E R
F Z B D E

The furniture selected by designer Pili Molina from Masfotogenica Interiorismo, consists of generally effective parts with lots of personality but doesn't take up extra spaces. The furniture is partly designed and manufactured, along with the selection of many details from a variety of shops and markets.

One of the most successful ideas was to visually unify the dining room with the hall by the same dark green color paint. In this way everything is integrated and the space is doubled. Mirrors distributed throughout the house have been greatly unified the interior and provid a double service: useful for future tenants while increasing the spatial feeling in a very interesting way.

家具是由Masfotogenica Interiorismo室内设计公司的设计师皮利·莫利纳精心挑选的，其中包含很多有个性的部分，但并没有占用多余的空间。有些家具是精心设计制造的，细节方面的物件也是从各种商店和市场挑选的。

这个项目最成功的想法是将餐厅和大厅刷成同样的深绿色，这使空间看起来大了两倍。分布于整个房子里的镜子使室内有机结合起来，并提供了双重服务：既有利于将来的住户，又以有趣的方式增强了空间感。

YOU
SOUL
VINTAGE
LONG TRACK

TINTIN
TINTIN
CONGO
SOUL
VINTAGE
FREE
75
LONG TRACK

AESTHETICS OF LIFE

生活美学

设计公司

DESIGN COMPANY

Interior Design Studio Marta De La Rica

设计师

DESIGNER

Marta De La Rica

摄影师

PHOTOGRAPHER

Pablo Zuloaga

This project could probably be the closest to perfectly reflect Marta De La Rica's style, decorative taste and personal design ideas. The apartment is located in a centric, modern, dynamic, young and cheerful neighborhood in Madrid called "Justicia". The building was built in year 1905 and since that date it had never gone through any major works.

这个项目近乎完美地反映了Marta De La Rica室内设计公司的风格、装修品位及个体设计灵感。该公寓坐落于Justicia，是马德里市中心附近一个时尚、动感、年轻、欢快的地方。该建筑于1905年落成，从那以后，再没有经历过任何大型工程。

When the designer Marta De La Rica and her husband bought the house, it was an authentic ruin. However, it had plenty of elements that captivated her: carpentries, the height of the roofs, the existing floors..., but above all, the light. Once they started the works, even if they tried to maintain as much as possible, they realized that many of the existing elements could not be reused. This is why they made lots of efforts thinking of ways to maintain the house's soul and did a hard restoration work.

Their first goal was to be able to reach a modern distribution according to their needs and moment in life. Marta suggested an easy and comfortable circulation, as well as a linear visual effect letting all the spaces been seeing from one side to another of the house. This line was also characterized by the rhythm settled by the seven balconies pointing to the street. This way, Marta decided to design the kitchen separated from the dining room only by a metallic and window screen with a single hung window and a hidden shade in case there was the need of separating spaces. The dining room was separated from the living room by a double door thought to be opened most of the time. The living room was divided into two different spaces by a metallic structure she designed, behind which you could find a working space that could also function as the main bedroom hall. Corridors were avoided all over the apartment giving way to the different rooms through the original doors restored and relocated.

当设计师Marta De La Rica和她的丈夫买下这所房子时，它已是一片废墟。然而，还是有一些元素吸引了她：木作、屋顶高度、现存的地面……但最重要的是，光。当他们开始动工的时候，尽管尽力保留，还是发现很多现存的元素不能被重新利用。这就是为何他们花费很大力气留存这所房子的灵魂并且进行了一场艰辛的修复工作。

他们的首要目标是根据日常生活需求进行现代化空间分配。Marta提出一种简单舒适的循环模式，以及线性视觉效果，让屋内所有空间都能从一边看到另外一边。这条主线的特点是七个朝向街边的阳台构成的韵律。Marta决定用一扇金属屏风隔离厨房和餐厅，悬窗和隐藏的遮蔽物是为了分离空间。餐厅及客厅经由一扇双开门隔开，那扇门大多数时间是打开的。她打造了一个金属结构体，在客厅中隔开两个不同的空间，你可以在后面找到工作的地方，同时可作为卧室使用。整套公寓都没有走廊，目的是通过修复和重置过的原始大门，为不同的屋子腾出空间。

As Marta mentioned before, she tried to reuse as many elements as possible, like the original doors, of course restored and in most cases, relocated. The floor she chose for almost the whole apartment was made out of robust pine wood from Denmark (Dinesen is the brand). The truth is installing the floor became an authentic odyssey, from its transportation to its laying. However, the beauty of this natural, long and width floor made all these efforts more than worth it.

The kitchen and bathrooms floors were tiled with special hydraulics recalling the original flooring. Walls and roofs were conceived in order to form a big white box, supported by the clarity of the floor and even more emphasized by the light entering through the windows. This white box was completed with walls playing with different textures, colors and aspects. This way, for example, the chimney wall in the main bedroom was covered by purple velvet that not also contrasted with the painting hanged in it but it also emphasized the end of the visual line from the kitchen to the main bedroom.

Marta曾经提过，她试图尽量重新利用那些元素，比如说最初的大门，当然，大多数都得以修复和重置。几乎整套公寓的地板都由丹麦（Dinesen品牌）结实的松木制成。事实上，从运输到铺设，安装这种地板成为一次真正的冒险之旅。然而，它的自然、修长与宽阔之美令所有努力变得物超所值。

厨房和浴室的地板采用了特殊的液压技术，令人回忆起最初的样子。墙壁和屋顶被设计成一个白色大盒子状，地板的明晰和窗外透进来的光都衬托了它的美感。这个白色盒子的建成，有赖于不同的材质、色彩以及朝向。比如，主卧室的烟囱墙由紫色丝绒覆盖，不仅与房内的油画形成对比，而且强调了从厨房到主卧室的视线轴端点。

TABELLÆ
ALUMEN
USTUM
SULFUR

Because of Marta's profession and above all, her passion, she has been buying unique objects she liked, since even before planning to have her own house. As a consequence, it was her opportunity to place them. All these objects are the result of many days and hours of finding, many excursions and great enjoying moments doing what she loves. Various things were bought during and after the works in the apartment and others, she hopes to keep finding later and changing ones for others. In the end, her goal is to create a combination of architecture, decoration, furniture, light, art, memories... where everything is mixed with naturalness, freshness and harmony. Fortunately, she can assure they made it!

因为Marta的专业，尤其是热情，甚至在她还没有想过买房时，就已经在购置自己喜欢的奇特物件了。因此，她有机会随心所欲的摆放这些物件。所有物件都花费很多时间寻得，期间她经历多次短途旅行并享受她所喜爱的精彩片刻。公寓装修前后，她购买了不同种类的东西，并且希望继续寻找或更换。最终，她的目的是打造一个建筑、装饰、家具、光、艺术和回忆兼容的地方，这里的一切都显得自然、新鲜与和谐。幸运的是，她确信它们实现了！

TREASURE TROVE OF MEMORIES

珍贵记忆

设计公司

DESIGN COMPANY

Andina & Tapia

设计师

DESIGNERS

Monica Andina & Fernando Tapia

摄影师

PHOTOGRAPHER

Rafa Dieguez for Nuevo Estilo/
Hearst Magazines Spain

HISTORIA DEL ARTE DE ESPAÑA
New York 1960
GREAT MODERN MASTERS
MEUBLES
IN THE PINK

This project was very personal for the studio, being Fernando´s apartment gave it a different approach to the usual projects the designers developed. It had to convey Fernando's personal touch above all, but also the way other designers do things in the studio.

这个项目对于这个设计公司来说是非常私人的，它是设计师费尔南多的公寓，与设计师们之前设计的项目有所不同。这个项目首先要考虑费尔南多个人的设计手法，其次是工作室其他设计师的做事方式。

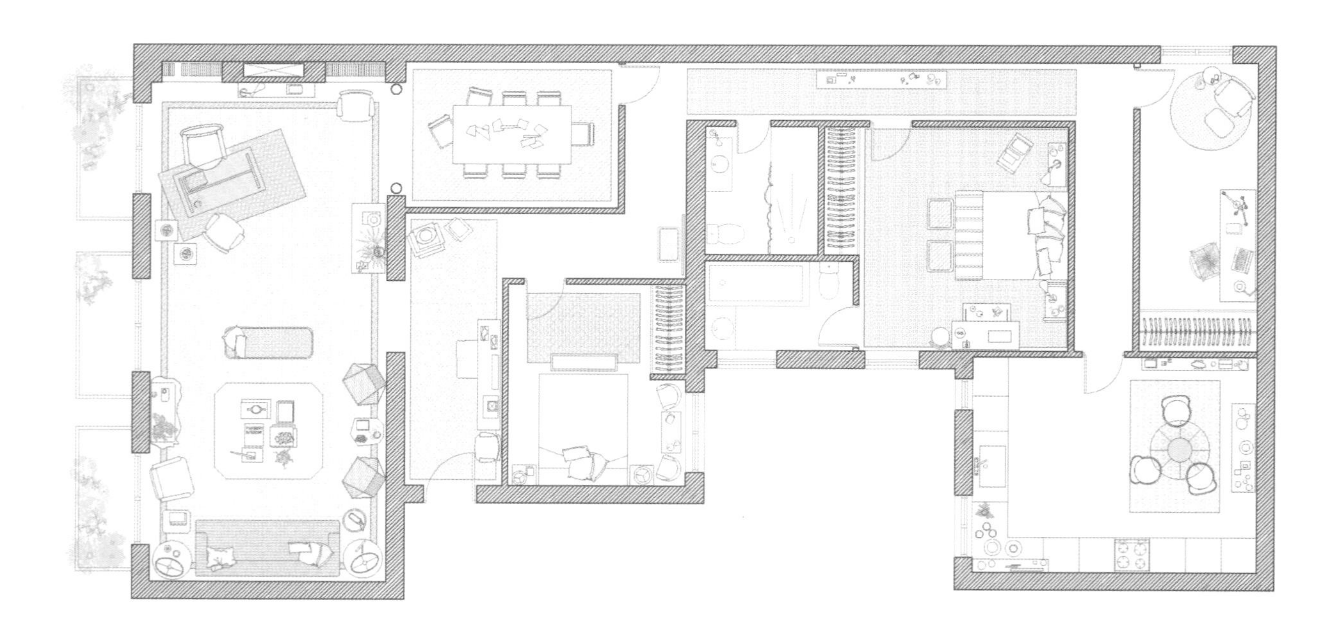

PARISIAN INTERIORS

The apartment had been renovated by the owner and the designers could not make a big renovation, all the original elements had been destroyed (except the 2 columns that separte living room from dining room), so they had to adapt to the space showing its virtues and hiding it´s defects only by painting the house and doing some works with the lighting. The project focused on creating a comfortable space for a daily basis and entertaining friends. The color scheme is very neutral with different tones of reds throughout the house, except the master bedroom in yellows.

公寓已经被之前的业主翻新过，因此设计师们不能进行大改造。除了分隔客厅和餐厅的两个大柱子，原来所有的元素都被毁坏了，所以设计师们只能适应这个空间，通过粉刷和光线来展示它的优点，掩盖它的缺点。这个项目专注于创造一个日常生活所需的舒适空间，也能招待朋友们。除了主卧里的黄色，整个房子都是以红色为主色调。

FAR FROM THE MADDING CROWD

远离城嚣

设计公司

DESIGN COMPANY

Andina & Tapia

设计师

DESIGNERS

Monica Andina and Fernando Tapia

摄影师

PHOTOGRAPHER

Pablo Zuloaga for Nuevo Estilo/ Hearst Magazines Spain

The house built in the 40´s feels like a small townhouse in the middle of Madrid, very close to the bullring. It is divided into four floors and has approximately 270 square meters. The original layout was interesting, with open spaces full of light, but the furniture was poor and the colors of the woodworks were not appropriate. So the designers focused on enhancing its virtues. With this idea in mind, although the adjustment was small, there was a great change. Only the below deck and the kitchen were completely redone.

这个房子看起来像个小的联排别墅，始建于40年代，位于马德里的中心，离斗牛场很近，分为四层，有大约270平方米。

最初的格局是很有趣的，空间开放，光线明亮。但家具很旧，木制品的颜色很不合适。所以设计师们决定集中精力提升其优点。基于这个想法，尽管他们对空间只做了微调，整个变化还是很大的。只有地板和厨房全部重新设计了。

CASA
RENOIR
LOFTS
Leonardo da Vinci
ABCDF

When entering the house what catches your eyes at first sight is the patio, with its flowers and trees, that's the moment when you live behind the bustle of the city. The designers' idea was to maintain that idea of stillness and escape the city's daily speed. The warm materials on the top of the stone floors and the furniture in textiles as with some pieces of furniture with a country side influence, make the house somehow trap you inside.

The designers tried to enhance the virtues that gave the house its own identity with a warm and low lightning, lots of fabrics, noble and simple materials. They wanted to give the owner open and big spaces to enjoy with the family and friends. But even being open spaces, each member of the family should be able to find its own space.

On the ground floor, the designers find the public spaces. From the entryway that embraces you in a darker space, you get access to the main space very big in which it has unified kitchen, dining room, living room and a small working area, all avoiding the use of doors or walls.

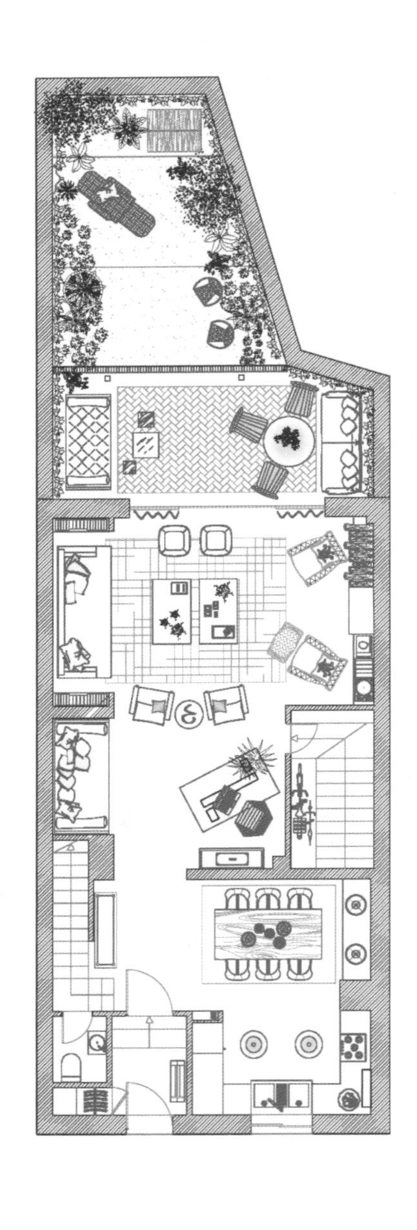

Cartier

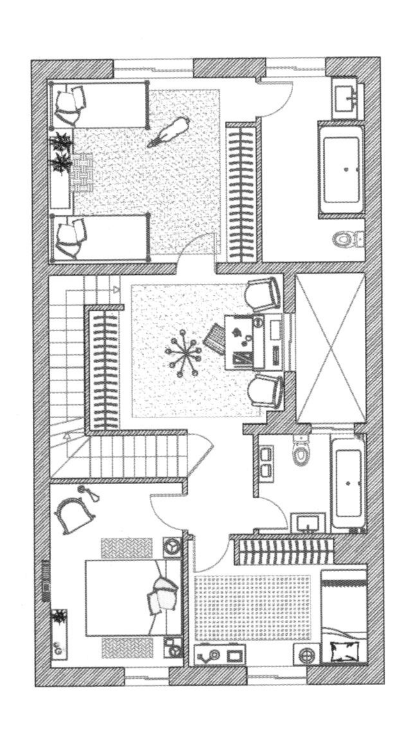

走进房子，首先映入眼帘的是露天庭院，有花有树，在这里，你可以享受远离城市喧嚣的美好时光。设计师的理念就是保持宁静和远离城市的繁忙。石头地板上的毛毯，纺织品家具以及一些有乡村气息的家具，都会让你不知不觉陶醉在这个房子里。

通过温暖的光线，大量的纺织品，简单而不失高贵的物品的装饰，设计师们提升了这个房子的优点和辨识度。他们想为屋主提供开放的大空间以便与家人朋友共享，即使是开放的空间，家庭的每个成员也有自己的私人空间。

在一楼，设计师们找到了公共区域。从黑暗的入口通道走进来，进入一个大空间，这里有厨房、餐厅、起居室和小面积的办公区，没有门和墙壁，所有的都在一个大空间中。

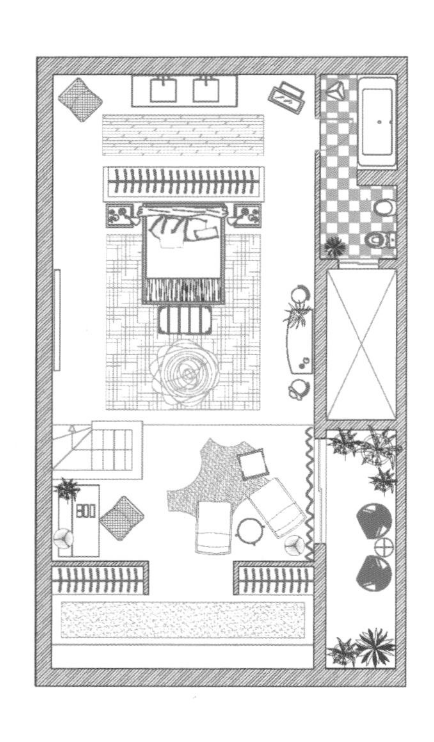

MODERN AND SOPHISTICATED

现代精致

设计公司
DESIGN COMPANY
Andina & Tapia

设计师
DESIGNERS
Monica Andina & Fernando Tapia

摄影师
PHOTOGRAPHER
Belen Imaz

The 250 square meters space had previously been an office, for that reason the distribution had lost the essence of a home. The challenge to the designers was to design a home for a family with four children and their specific requirements. The most important criteria when solving the distribution were to create a convenient space, recovering the essence of a nineteenth century building and recreating some of its architectural elements but always with a new contemporary approach.

这个250平方米的空间之前是个办公室，因此它的分布不同于住宅的分布。设计师们所面临的挑战是设计一个适合有四个孩子的家庭居住的家并满足他们特定的需求。解决分布问题最重要的标准是创造一个方便的空间，还原这个19世纪建筑的本质，重建某些架构元素，这些都基于新的现代设计方法。

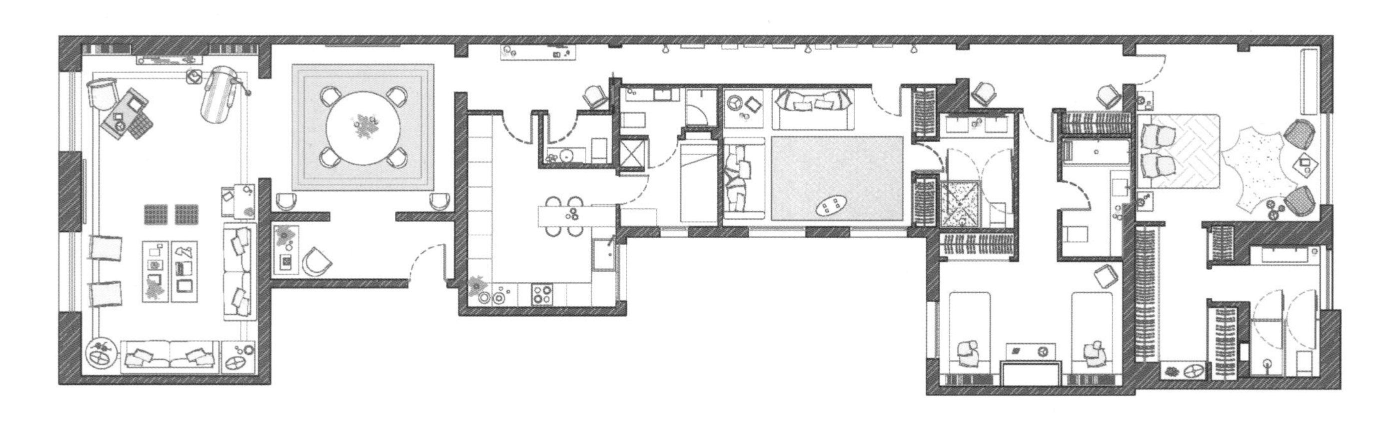

The client wanted a unified look and feel for the apartment, so the designers decided to use the same element on floors and bathroom coatings. Once all the restoration was made the designers also had to accomplish the decoration but they did not take a unique style. They tried to give the space a modern and sophisticated atmosphere, using furniture from different periods and styles and making them be together harmoniously.
When entering the apartment you can see a small entryway, created with a combination of mirrors and an antique chair with a contemporary cabinet at the end. The entryway leads to a dining room area. It looks like the continuation of the entryway with a big dining table designed by the designers, inspired by a Jansen model of the 70´s.
In the living room, there are two big windows with views to the park, two iron libraries designed by Andina and Tapia and an old fireplace in the center they recreated. The space is divided into a working area and a sofa area where the coffee table was also designed by them. The four bedrooms have their own individual bathrooms. All rooms with different approaches in terms of color schemes.

客户想要一个外观和感觉统一的公寓，所以设计师们决定地板和浴室用相同的涂料。当所有的修整都实现后，设计师们还必须完成装饰，但他们并没有采取唯一的风格。通过将不同时期和风格的家具和谐地摆放起来，设计师们给这个空间营造了一种现代精致的氛围。

走进公寓，你可以看到一个小的入口通道，这个通道是用镜子做成的，里面有一把款式古老的椅子，尽头处有一个现代橱柜。入口通道直通用餐区，用餐区有一张很大的餐桌，这个餐桌是由设计师们设计的，灵感来源于70年代的詹森模型。整个用餐区看起来像是入口通道的延续。

客厅有两扇大窗户可以看到公园的景色，还有两个由设计师安迪纳和塔皮亚设计的铁制书橱和一个古老的壁炉。整个空间分为工作区和休息区，休息区的咖啡桌也是他们设计的。四个卧室都有独立的浴室，每个房间的配色方案也是不同的。

L'AIR
MAPA
HOME

A TRANQUIL ATTIC

静居

设计公司
DESIGN COMPANY
Inés Benavides

设计师
DESIGNER
Inés Benavides

摄影师
PHOTOGRAPHER
Germán Saiz

The main idea was to take maximum advantage of the views and light coming from both sides of the flat. Designers, therefore, decided to place a living-room in each side; the first one faces an important monument in Madrid, and the second is opened to a beautiful and quiet terrace.

在本案中，设计师的主要意图是最大限度地利用公寓两旁的光线及视野。基于这一理念，设计师决定在公寓两侧各设一间客厅，其中一间面朝马德里的一个重要的纪念碑，另一间朝向一个安静祥和的露台。

At the back, the living-room belongs to the master bedroom but can be separated from it with a panel. The centre of the flat does not have as much light as the front and the back. For this reason,they opened interior windows between the rooms. They are not only decorative, but also they help the light to spread all around.

在公寓的后侧，护墙板可将原本属于主卧室的客厅分隔开来。

公寓中心的光线不如它前后那么亮，因此，设计师们在房间之间设计了大型窗户，这不仅起到了装饰的作用，而且可以使光线扩散到公寓四周。

This flat is located in a XIXth century building in the centre of Madrid, and it was important to keep the spirit of the original building. The flat is very contemporary and even if designers rebuilt it from scratch the original elements are still present.

The decoration is bright and vital but quality and elegance had to be the most important characteristics. To achieve this, designers chose open spaces, noble materials and contemporary designs for furniture.

The best part is the terrace, so designers tried to integrate it as much as possible to the flat. It makes the spaces more lively and human.

这间公寓坐落于马德里市中心的一栋19世纪的建筑物内，它对于保持这栋建筑物原本的韵味而言起着重要的作用。这间公寓的设计偏现代化，即便设计师们对它进行重新装修，公寓仍保留着它原本的韵味。

它的整体装饰给人明亮、有活力的感觉，但又并未忽略质量和高雅这两个重要的特点。为了达到这一效果，设计师们选取了开放空间，采用了优质材料和现代的家具设计手法。

露台的设计是这间公寓的点睛之笔，设计师尝试着将它与公寓融合，从而使公寓具活力和人文色彩。

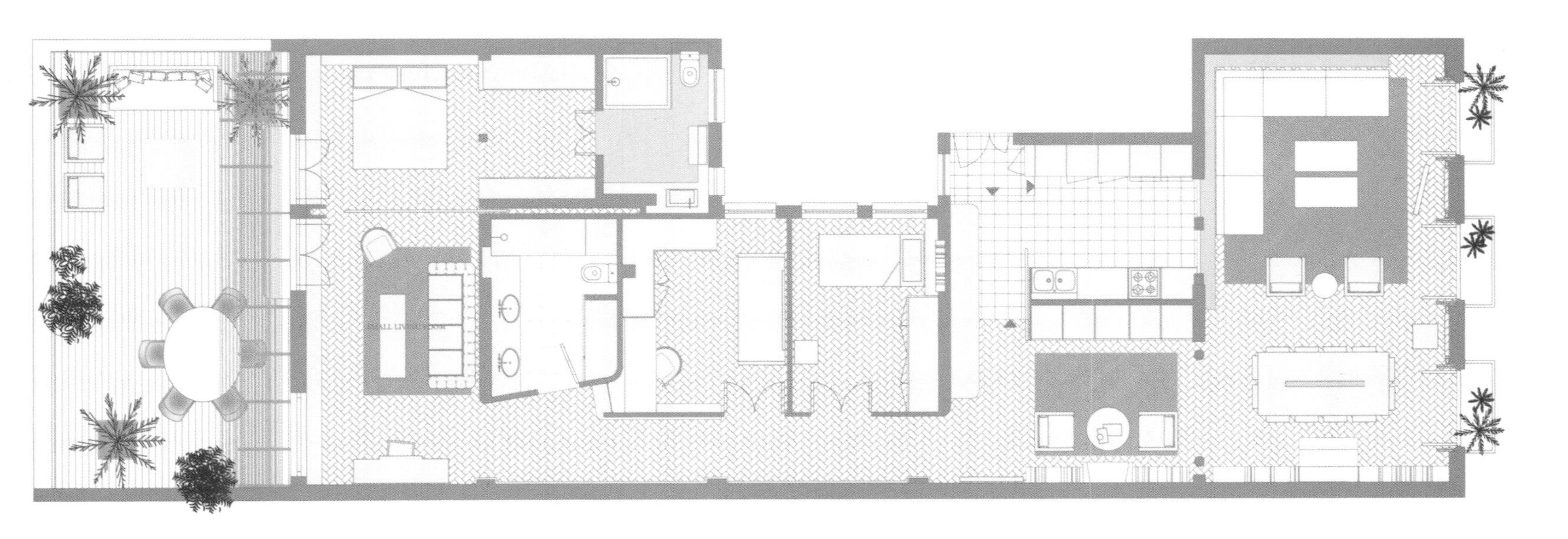

COMFORTABLE AND COZY

舒适惬意

设计公司
DESIGN COMPANY
Kayserstudio

设计师
DESIGNER
Georg Kayser

摄影师
PHOTOGRAPHER
Roberto Ruiz

The designer has been fascinated with the bohemian area of Gracia and chose to live and work in this area.

The apartment is set within a 1920´s building. It is built in the neoclassical style, and reflects perfectly the spirit of the city. Its many special and unique details, like the decorative mosaic tiles, or the arched doors, formed the prefect base to work with. But the most impressive feature is the large bay window, from where you can see over Barcelona until Montjuic.

设计师曾对格雷西亚的波西米亚地区心驰神往，便选择在此生活和工作。

该公寓位于20世纪20年代的一栋建筑物中，采用新古典风格，完美映射出这个城市的精神。很多独特的细节，比如装饰性马赛克瓷砖或者拱形门，形成了很好的工作基础。然而给人印象最深刻的特色是大型飘窗，从那里，你可以看到巴塞罗纳直到蒙特惠奇山。

The distribution of the apartment itself is very typical for the houses of the Eixample, the newer part of Barcelona, which was built when the city started to grow out of its medieval walls. The designer tried to respect the original distribution as much as possible. Accessing the apartment from the center, a long and wide corridor connects all rooms. As the designer wanted to maintain the distribution and elements, the designer mainly used color to create a more contemporary feeling: a pale blue in harmony with the blue of the mosaic tiles was selected for the corridor. Dark beige was chosen for the living room, and a pale green for the bedroom. The rest has been painted in a warm off white.

The selection of the furniture and decoration is an eclectic mix of different styles and époques. Danish modern furniture mingles with antique vases and vessels; rustic French pieces are juxtaposed against the contemporary art. But everywhere—in the mosaic-tile floors and the traditional wall moldings—the memory of the old remains.

该公寓布局在巴塞罗纳新扩展区房屋中非常典型，建于该市开始摒弃仿中世纪墙壁设计之时。设计师设法尽量维系原始布局。从中心进入公寓，有一条长而宽阔的走廊连接各个房间。因为设计师想要维持布局元素，他仅用色彩来打造更现代化的感受：走廊采用与马赛克瓷砖的蓝色协调一致的灰蓝色，客厅选用了暗米色，卧室则运用灰绿色，其余房间采用温暖的米白色。

家具和装饰物的选择是不同风格、不同时代的折衷混搭。丹麦现代家具与古董花瓶及容器混合；淳朴法式物件与当代艺术并列。然而随处可见的马赛克瓷砖地面与传统墙壁线脚，都是遗留下的古老记忆。

MfS AOP 1010/91 Bd. 13 BStU 000202
CALABRE

CALABRE

To tone down the bright sunlight that floods the living room during the day, the designer opted for masculine color palette that includes muted gray walls and partly dark furnishings. A glass cocktail table and vacant frames keep the room feeling airy rather than overly moody. A voluminous hand-loomed rug by Barcelona producer Nami Marquina adds warmth competing with the intricately tiled original flooring.

The designer wanted to create relationships among seemingly disparate items from various eras. For example in the dining area, a worm French workbench purchased at a local flea market counters a sleek chrome table lamp, a lamp by Catalan designer Miguel Mila, and a 1920's clock from Germany.

The clash of époques continues in the kitchen. The designer left the cupboard for the plates and glasses, and installed a new kitchen, where the designer incorporated the old marble sink. A shiny white 1950's sculptural round table bought in the France contrasts with a pair of wooden chairs from Sweden.

为缓和白天直射入客厅的明媚日光，设计师选择了阳刚配色，包括柔和的灰色墙壁以及局部深色的陈设。玻璃鸡尾酒台与空框架使这里看起来轻快，而不是过度情绪化。巴塞罗纳制作者Nami Marquina的手织地毯与精致瓷砖铺就的原始地面一较高下，增添了温馨之感。

设计师希望用不同时代看似不相干的东西打造出一些关联。例如，餐厅中购于当地跳蚤市场的法式工作台与光滑的加泰罗尼亚设计师Miguel Mila所做的铬黄桌灯以及一座20世纪20年代德产钟产生对立。

时代的碰撞在厨房延续。设计师留下橱柜放置盘子和杯子，并设置了一间新的厨房，包含古老的大理石水槽。20世纪50年代的光面白色雕刻圆桌带入了法式风格，与瑞典的木制座椅形成对比。

51

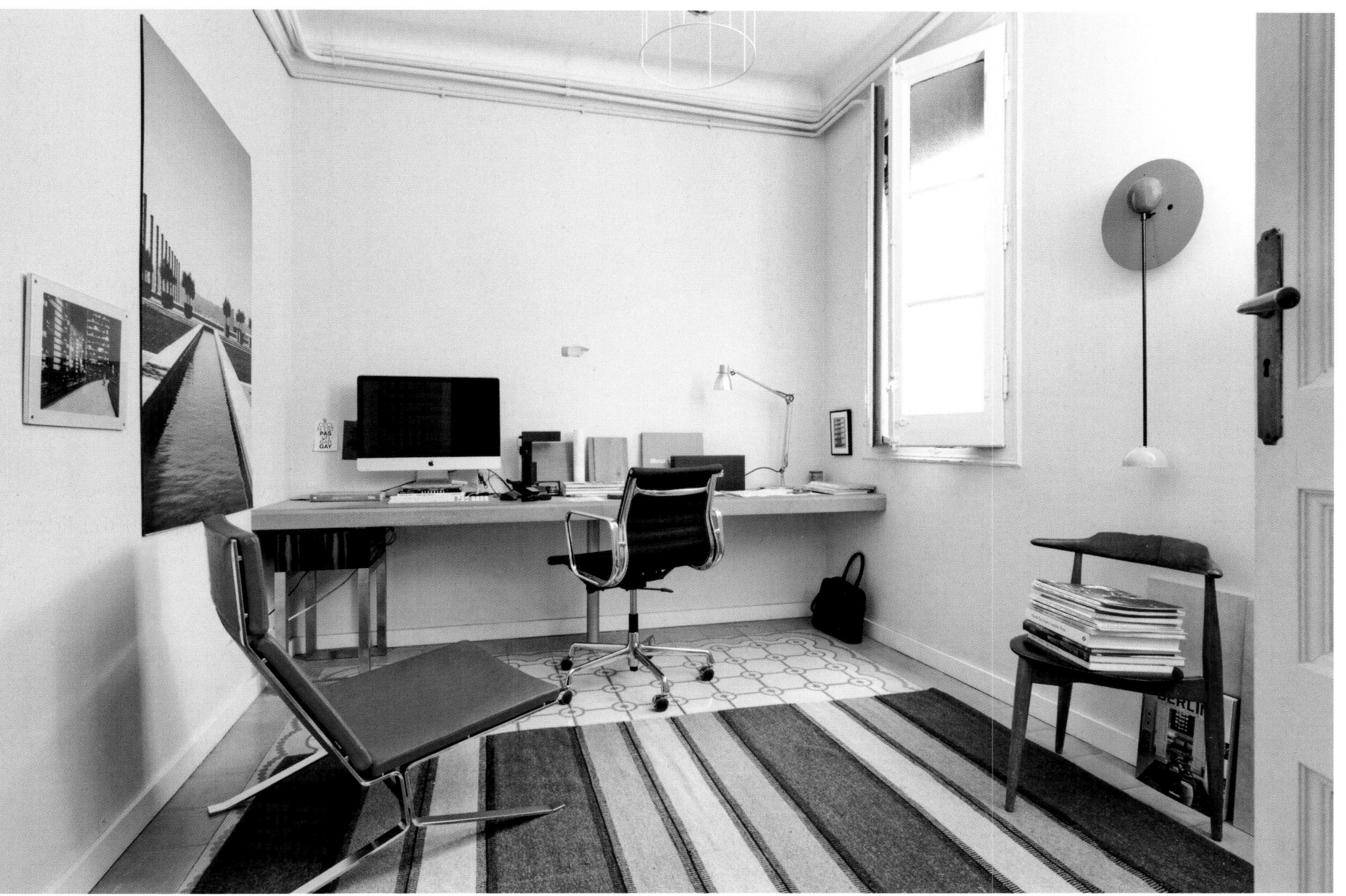

never read.
pictures.
Andy Warhol

TIVOLI

TIVOLI

The main bathroom received a new flooring and partial tiling designer mixed it with an old bathtub, and the original sink.

主卫铺砌了新地板以及部分瓷砖。同样，这里也有古老的浴缸和原有水槽的混合搭配。

A JOURNEY OF ART

艺术之旅

设计师

DESIGNER

Nuria Mora

摄影师

PHOTOGRAPHERS

Manolo Yllera, Amaya de Toledo

Gunta Stölzl

Nuria Mora is a Spanish artist, she started painting in the streets of Madrid in 1999.
She has traveled around the world painting in unusual places and create ephemeral art.
Surprised to find her drawings in a few doors from the favelas of Rio de Janeiro, on the beaches of Mexico, the streets of Cuba, Chile, Argentina ...
Her house, a penthouse loft renovated by her is located in the neighborhood of Lavapies in Madrid, reflects her personality, spontaneous but not casual. Everything is visible, exposed as a museum of the senses: the brushes and pencils with drawing , and all objects and household items, the most delicate English Royal Albert porcelain and Mexican clay dishes.

纽莉娅·莫拉是一个西班牙艺术家，她1999年在西班牙街头开启了她的绘画之旅。

她周游世界，在一些不同寻常的地区进行绘画创作，留下短暂的艺术品。

人们可以惊奇地在里约热内卢的一些平民窟的门上，墨西哥的海滩上，古巴、智利、阿根廷的街头等地方发现她创作的绘画。

她自己设计的复式阁楼毗邻马德里的拉瓦皮埃斯区，这间阁楼的设计反映出这位艺术家与众不同的性格。在这间阁楼里，一切都是清晰可见的，人们可以看到里面像博物馆似的陈列着的物品。比如绘画的刷子和铅笔，一些家庭用品, 极为精致的英国皇家阿尔伯特瓷器和墨西哥陶瓷碗。

To her, the aesthetic is important, prioritized over profit .The idea was not to make a loft as a concept, but an open space that would have views.

对设计师而言，审美尤为重要，它的重要性超过了利益。设计师的理念不是使“阁楼”只存在于概念中，而是将它设计成一个可供欣赏的开放空间。

Audubon's Birds of America

ELEGANT AND MAJESTIC

尊贵典雅

设计公司
DESIGN COMPANY
SERRANO SUÑER ARQUITECTURA

设计师
DESIGNER
Fernando Serrano-Súñer de Hoyos

摄影师
PHOTOGRAPHER
LUZESTUDIO

For this project the client has opted for a wooden floor Oak European dimensions 290mm x 20mm x 3000mm with the peculiarity that the quality of the wood is select (without any knots) certainly a choice that brings elegance and majesty to the already huge rooms of this house.

All steps of this house are also designed and made with the same wood used for the floor, so although each has a different design stairs all bear the same tune as the floor.

Certainly for the designer the tone of this antic oak contrasts perfectly with the light tones of walls and carpentry inside giving it a lot of warmth to the rooms.

To the designer this project conveys a feeling of home.

本案客户选择了尺寸290毫米 x 20毫米 x 3000毫米的欧洲橡木地板，其木料质地的特性为房内巨大的房间带来典雅与威严并存之感。

所有台阶都使用了与地板相同的木料，尽管设计上各有不同，但仍然具有与地面一样的调性。

当然，对于设计师来说，这种橡木调性与墙壁和木工的浅色调形成了完美对比，为各个房间提供了许多温馨之感。

于设计师而言，本案传递了家的感觉。

PROVINCIA DE HUELVA
PROVINCIA DE SEVILLA

SPAIN LIVING · MADRID

COZY LIVING

惬意生活

设计师

DESIGNER

Dafne Vijande

摄影师

PHOTOGRAPHER

Santiago Moreno

This house is designed for a large family and it has different spaces for all members of the family. The fundamental aesthetic is the creation of an atmosphere of clarity with small touches of color. Here light plays a very important role in all the spaces as they should be very functional as well as aesthetic. The result is very harmonious. The house is distributed into several floors with common areas on the main floor. The living room has several areas: work, relax and reading area. Pictures play a fundamental role because they have lots of personality.

此案是为一个大家庭设计的，设计师为每个家庭成员都设计了不同的空间。设计师的基本设计理念是在多彩的空间中创造一个清新的氛围。吊灯在所有空间中都起着非常重要的作用，不仅起到照明作用，也带来审美体验，使空间更加和谐。此案为多层设计，主要楼层都有公共区域。客厅分为三个区域，即工作、休闲和阅读区。墙上的油画起着重要的作用，因为它们可以彰显个性。

the relaxed home Atlanta Bartlett
Juegos de agua
Carlo Scarpa
TASCHEN
Casas norteamericanas
BE
GHOST TOWNS
OF ARIZONA
TIZAS
DADOS
PELOTAS
BEBÉS
COCHES
LÁPICES
FICHAS
PELUCHES
A
PICASSO CLÁSICO
RAMON GAYA
En la mesa de

The kitchen has several areas, with the dining room included in it that has a lot of natural light because the family spends there many hours a day. The master bedroom is neutral with soft colors and it has a dressing area. The children's bedrooms are more colorful, according to the tastes of children and each one has its own bedroom.

The exteriors are designed with colors to give an atmosphere of fantasy. In addition, there is a garden that children begin to get in touch with nature. The terrace area is created for the elderly with a relaxation area.

厨房分为多个区域，其中包括餐厅，餐厅光线充足，家庭成员每天都在那里度过很多愉快的时光。卧室采用柔和的色调，包括一个更衣区。儿童房颜色鲜艳，这是根据他们自己的品味设计的，每个孩子都有自己独立的卧室。

室外的设计也选用了鲜艳的颜色，以便于营造一种梦幻般的氛围。此外，孩子们可以在花园玩耍，亲近自然。露台区的设计是为了使年迈的人在此放松心情。

FURNITURE
SOUTH AFRICAN STYLE

PAPÁ
Tristán
León